Return to George Stimson

FIRST® ROBOTS
✹ BEHIND THE DESIGN
30 Profiles of Design, Manufacturing, and Control

© 2015 by United States Foundation for Inspiration and Recognition of Science and Technology (FIRST®)

All rights reserved. No part of this book may be reproduced in any form without written permission of the copyright owners. All images in this book have been reproduced with the knowledge and prior consent of the artists concerned, and no responsibility is accepted by producer, publisher, or printer for any infringement of copyright or otherwise, arising from the contents of this publication. Every effort has been made to ensure that credits accurately comply with information supplied. We apologize for any inaccuracies that may have occurred and will resolve inaccurate or missing information in a subsequent reprinting of the book.

FIRST®, the FIRST® logo, FIRST® Robotics Competition, FRC®, FIRST® Tech Challenge, and RECYCLE RUSH℠ are trademarks of FIRST®. LEGO® is a registered trademark of the LEGO Group. FIRST® LEGO® League, FLL®, Junior FIRST® LEGO® League, and Jr.FLL® are jointly held trademarks of FIRST and the LEGO Group. All other trademarks are the property of their respective owners. ©2015 FIRST. All rights reserved.

First published in the United States of America by
FIRST®
200 Bedford Street
Manchester, New Hampshire 03101
Telephone: (603) 666-3906
firstinspires.org

Library of Congress Cataloging-in-Publication Data available

ISBN 13: 978-0-692-54062-6

10 9 8 7 6 5 4 3 2 1

Design: Kristen Stutt, NFUSION, Inc.

Cover Image: FIRST® Robotics Competition Team 118 Robot

Photo Credits: Each profiled team supplied the images that appear in this book. Additional image credits include Wesley Schaefer (pg. 8), FRC Team 1692 (pg. 21), Dan Donovan, courtesy of FIRST® (pg. 61), Lilly Kam/FRC Team 4964 (pg. 97), Miriam Somero, courtesy of FIRST® (pg. 129), Argenis Apolinario (pg. 131), Ronnie Ann Tront (pg. 171), Lou Ravetta, courtesy of FIRST® (pg. 209), and Adriana M. Groisman, courtesy of FIRST® (pg. 246).

Printed in USA

FIRST® ROBOTS
⚙ BEHIND THE DESIGN
30 Profiles of Design, Manufacturing, and Control

Vince Wilczynski and Stephanie Slezycki

FOREWORD BY DEAN KAMEN
Founder, *FIRST®*

AFTERWORD BY WOODIE FLOWERS
FIRST® Distinguished Advisor

ALSO BY VINCE WILCZYNSKI
AND STEPHANIE SLEZYCKI

*FIRST Robots: Aim High
– Behind the Design*
©2007 BY ROCKPORT PUBLISHERS, INC.

*FIRST Robots: Rack 'n'
Roll – Behind the Design*
©2008 BY ROCKPORT PUBLISHERS, INC.

CONTENTS

 TABLE OF CONTENTS

FOREWORD 9

CHAPTER 1
Design, Manufacturing, and Control of Robots 10

Building Futures by Building Robots 12

The Power of FIRST® Mentors 20

CHAPTER 2
Computer-Aided Design, Simulation, and Analysis 22

Team 118 – Champion of Design 24

Team 846 – Monkeying Around with Gears 34

Team 1114 – Computer-Aided Design as One Step of the Design Process 40

Team 1511 – Many Uses of Computer-Aided Design 48

Team 1987 – The Broncobots' Conveyorbot 54

The FIRST® Robotics Competition Chairman's Award 60

CHAPTER 3
Innovative Design Using Traditional Machining Methods 62

Team 525 – Designing Simplicity 64

Team 2481 – Strategic Use of Machining Resources 72

Team 2485 – Spotlight on Carbon Fiber 78

Team 3339 – Achieving Excellence with Basic Tools and Design Ingenuity 82

Team 3478 – Robot Redesign: Doing a "180" 90

FIRST® Impact — Transforming a Community 96

CHAPTER 4
Three-Dimensional Printing 98

Team 125 – Carefully Planned Incorporation of 3D-Printed Components 100

Team 359 – Printing a Prototype 106

Team 2601 – Additive Manufacturing Inspiration 112

Team 3824 – Large Scale Additive Manufacturing of Robots 118

Team 5030 – Fast Solutions for Small Parts 124

FIRST® Dean's List 128

CHAPTER 5
Computer-Controlled Machining with Mills and Lathes 132

Team 236 – Mastering CNC Mills and Routers 134

Team 696 – The Circuit Breakers: CNC Extraordinaire 142

Team 1538 – The Holy Cows: Leading the Herd in Computer-Controlled Machining 150

Team 3250 – Efficiencies with Automated Manufacturing 158

Team 4293 – Custom Components for Competitive Robots 164

FIRST® Robotics Competition Makerspaces 170

CHAPTER 6
Computer-Controlled Cutting Systems 172

Team 67 – Heroes of Tomorrow 174

Team 148 – The Robowranglers: On the Cutting Edge of CNC Capability 180

Team 1983 – Above and Beyond with CNC Cutting Capability 188

Team 2848 – The Right Tools Enable Flexibility 194

Team 4183 – Wood 'bot, Good 'bot 202

Woodie Flowers Award 208

CHAPTER 7
Sensors, Monitoring, and Control Applications 210

Team 624 – Sensing Victory 212

Team 1100 – Simply Reliable 220

Team 2062 – Value of Simplicity and Reliability 226

Team 2168 – Custom Sensors, Software, and Driver Station Dashboard 232

Team 3310 – Control of an Industrial Arm 238

AFTERWORD 247

◀ Dean Kamen is President of DEKA Research & Development Corporation, a dynamic company focused on the development of revolutionary new technologies that span a diverse set of applications. As an inventor, physicist, and entrepreneur, Kamen has dedicated his life to developing technologies that help people lead better lives. Among Kamen's proudest accomplishments is founding FIRST.

FOREWORD
by Dean Kamen

FIRST® — An Experiment to Transform Our Culture

The concept of learning is an incredibly personal activity. Humanity is constantly provided with opportunities to solve the problems that surround us.

For the last 25 years, I have been involved with an experiment to focus on an area that deserves significant attention: the technical competency of society. This experiment was created to transform the culture of our society to one that values the contributions of scientists, engineers, and innovators who use their talents to improve our world. My hypothesis is that if students create solutions to challenging problems during their earliest years, they will be motivated to pursue higher education in engineering, technology, science, and other disciplines. The name of my experiment is FIRST® — a program that transforms our future, one person at a time.

The annual FIRST® Robotics Competition (FRC®) challenges teams of students and mentors to design and construct an electro-mechanical solution to a complex challenge. Each year the challenge takes the form of a "game" that must be played by robots — mechanisms that the teams conceive of, design, construct, program, and test in a period of only six weeks. Their solutions are then celebrated at competitions that draw thousands of cheering fans and supporters. In 2015, 2,900 FRC teams involved over 72,000 students and 28,000 mentors— plus legions of volunteers as well as corporate, government, and individual supporters. It is quite a large experiment, and the only one of its kind.

The process of learning on a FIRST team — while an individual process for each student — is a communal experience. Mentors are key in FIRST for they are the agents that transfer knowledge to the students. The learning that is achieved by each student involved in FIRST results from the guidance provided by FIRST mentors. That learning is not limited to the technical aspects of robot design, fabrication, assembly, and programming, but also extends into life skills such as communications, leadership, and teamwork. FIRST mentors come from every walk of life and are drawn together because of their commitment to improving the future. Though the purpose of FIRST and its focus on learning new skills centers around students, all members of a FIRST team ultimately learn from one another. Learning from others is a contagious component of FIRST.

This book looks at a narrow subset of the learning that takes place on a FIRST team. Design, manufacturing, and control technology has been rapidly advancing over the last decade and FIRST teams are immersed in those advancements. These examples illustrate some of the leading technical developments made by FIRST teams during the 2015 FRC season. Each chapter presents case studies from FIRST teams who excelled in aspects of the design, manufacturing, and control of their FIRST robot. Their results, and the inherent learning by the members of these teams, indicate that the FIRST experiment is a success.

It is amazing that all of the work that is showcased in this book was completed in only six weeks. FIRST teams were presented with a challenge to solve the first week of January and their solutions had to be completed by the second week of February. The progression from an idea to a completed device in six weeks, including prototyping, testing, and debugging, is inspiring. Doing so while both teaching and learning is even more inspirational. Taking the time to clearly document the creative process is extremely commendable.

My experiment to transform the culture using FIRST has benefitted from the collective work of individuals, corporations, businesses, schools, governments, and organizations. There is room for everyone in this experiment. The stakes are high and as a society this experiment cannot fail.

FIRST is much more than simply building robots — it is an effort to change our culture. Inspiration is the heart of FIRST, and that inspiration is multi-directional. FIRST students are inspired by their mentors and the mentors are in turn inspired by their students. This effect extends beyond the teams themselves as FIRST sponsors, supporters, and volunteers are also inspired by the success of FIRST teams.

And, I, too, am inspired.

If you want to be inspired, join us and participate in FIRST.

CHAPTER 1
DESIGN, MANUFACTURING, AND CONTROL OF ROBOTS

Similar to the information revolution that created instant access to data as a result of advancements in computers and digital communication, society is approaching a parallel development in the physical world: the arrival of the modern industrial revolution. The barriers to conceiving, producing, and controlling physical objects have been significantly reduced due to advancements in design software and manufacturing equipment. Traditionally these capabilities have been restricted to academic and industry research labs, but recent technology developments have extended these abilities to a much larger population.

Building Futures by Building Robots

The FIRST Robotics Competition challenges teams of students and their mentors to create ingenious systems to accomplish a complicated task. In 2015, teams acquired and stacked totes, containers, and pool noodles on raised platforms.

ADVANCING TECHNOLOGIES

An increased interest in rapidly creating physical objects has resulted from developments in three key technologies: design software, manufacturing equipment, and integrated control systems. Intuitive computer-aided design (CAD) software programs enable users to quickly master the basic functions needed to design sophisticated mechanical systems. Once designed, physical devices can be manufactured with a variety of traditional and modern machines including three-dimensional (3D) printers, laser and plasma cutters, and computer-controlled mills, lathes, and routers. Sensors that measure nearly any physical parameter can be easily integrated with microprocessor-based systems to monitor and control machine functions. Advancing from a sketch to a functioning object is a process in which more and more people, from a variety of backgrounds, are now participating.

This book explores the technologies associated with modern and traditional methods for design, manufacturing, and control of robotic systems. Case studies illustrate methodologies that have been applied to design, manufacture, and control robotic systems designed for the 2015 *FIRST* Robotics Competition (FRC®), the world's largest team-based program to interest high school students in pursuing careers in science, engineering, and technology.

FIRST ROBOTICS COMPETITION (FRC)

FIRST® (For Inspiration and Recognition of Science and Technology) was founded in 1989 by inventor Dean Kamen to inspire young people's interest and participation in science and technology. Based in Manchester, NH, the 501(c)(3) not-for-profit public charity inspires young people to be science and technology leaders by engaging them in exciting mentor-based programs that build science, technology, engineering, and math (STEM) skills; inspire innovation; and foster well-rounded life capabilities including self-confidence, communication, and leadership.

FIRST offers a progression of four programs, with leadership opportunities and hands-on experiences in robotics engineering and invention challenges for students between the ages of 6 and 18:

- *FIRST*® Robotics Competition (FRC®) for grades 9-12 (ages 14 to 18);
- *FIRST*® Tech Challenge (FTC®) for grades 7-12 (ages 12 to 18);
- *FIRST*® LEGO® League (FLL®) for grades 4-8 (ages 9 to 16, ages vary by country); and,
- Junior *FIRST*® LEGO® League (Jr.FLL®) for grades K-3 (ages 6 to 9).

Young people can join the international, after school STEM programs at any level. *FIRST* participation is proven to encourage students to pursue education and careers in STEM-related fields, inspire them to become leaders and innovators, and enhance their 21st century work-life skills.

The premise for all *FIRST* programs is to challenge teams composed of youth members and adult mentors to solve technical problems. Robotics is used as an engaging platform for the majority of *FIRST* programs. The FRC teams design and build robots capable of playing a mechanical version of a sport. In *FIRST* Robotics Competitions, robots take on the role of an athlete participating in a unique contest with and against other robots. Annually, FRC teams are presented with a new game

Points were awarded based on the number of stacked totes positioned on scoring platforms, up to a limit of six totes per stack, with extra points earned for containers and pool noodles placed on top of the totes.

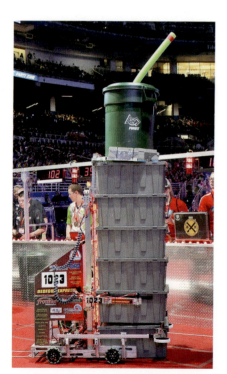

Driving skill combined with clever design was required to maneuver precious cargo and deliver each load to a scoring platform.

challenge the first week of January and must have their robots completed by the second week of February. With a new competition each year, the playing field between returning and rookie teams is somewhat leveled with regard to previous experience for each year's new game.

Following the six-week build period, the competition season consists of regional gatherings of 20 to 60 FRC teams that participate in a multi-day event where their designs are tested by competing with other teams. These events are celebrations of the technical achievements of each team, with awards presented for creativity, entrepreneurship, design, quality, team spirit, and other attributes. The competition season culminates in the FIRST Championship in St. Louis, MO, during April. In 2015, 2,900 teams from across the globe participated in the FIRST Robotics Competition and 607 FRC teams attended the FIRST Championship event.

A detailed set of rules regulates the components that can be used to build an FRC robot, as well as the maximum size and weight of the robot. Robots are limited to a 28-inch by 42-inch base but can extend unlimited in either direction while competing. Given this footprint and a maximum height and weight of 78 inches and 120 pounds, the robots end up being large and sophisticated devices.

In addition to the competition rules, FRC teams are provided with a kit of parts consisting of motors, electronics, mechanical hardware, software, and structural elements to use as a basis for their designs. Additional components as specified in the competition rules can also be used to build the robots. Also included in the kit of parts are computer-aided design (CAD) software programs provided by PTC, Autodesk, and SolidWorks — three of the leading CAD software companies. Teams use the software tools to design their robots, analyze performance attributes, and produce manufacturing plans.

A control system consisting of a NI roboRIO robot controller — a portable reconfigurable input/output (RIO) device — and other electronic control components are also included in the kit of parts. The roboRIO robot controller uses commands from the robot driver, as well as analog and digital inputs from installed sensors, to actuate motors and pneumatic systems on the robot. The controller is programmed in LabVIEW (a graphics-based programming language created by National Instruments), Java, and C++ to direct and automate functions on the robot.

2015 FIRST ROBOTICS COMPETITION: RECYCLE RUSH℠

The 2015 FRC challenge was titled RECYCLE RUSH℠ and required an alliance of three robots to work together to stack totes on a scoring platform and cap the stacked totes with recycling containers. Pool noodles, representing litter, placed in the recycling containers earned the alliance additional points. Stacks could include up to six totes, with the stacks earning points when placed at two locations within the 26-foot by 27-foot playing field allocated to each of the two alliances competing during the same time period. Additional

Each match started with an autonomous period where robots operated based on programmed instructions to collect and stack yellow "autonomous totes." Points were awarded for stacks placed in a specified field location.

points were awarded by cooperating with the alliance on the other half of the playing field to stack four specially designated yellow totes on the center step. Robots on one alliance could not interfere with the operation of robots on the other alliance, with the exception of interactions associated with obtaining the containers in the center of the field.

The competition included autonomous and teleoperated periods of play with a fifteen-second autonomous period occurring at the start of each match.

During the autonomous period of play the robots performed without real-time input from the drivers. Instead they relied on sensor feedback to execute programmed algorithms. The primary goal for each alliance during this period of play was to stack three yellow totes in a designated location on the field. With a field setup of alternating yellow totes and upright containers, navigation and robot manipulation techniques were required to acquire and stack the yellow totes without input from the drivers.

A secondary alliance goal during the autonomous period was the acquisition of the four recycling containers at the center of the field, a location accessible to both alliances. Given that the value of a stack of totes was tripled when capped with a recycling container, obtaining as many containers as possible increased an alliance's scoring potential. Robots raced one another to obtain the center containers at the start of the autonomous period by grabbing the containers and dragging them to their side of the field.

Team members interacted directly with robots at each corner of the playing field. In these designated areas human players loaded robots with totes and pool noodles that earned points when deposited in a container. Human players could also throw the pool noodles on the opposing alliance's field to garner extra points.

16 | FIRST Robots: Behind the Design | Vince Wilczynski and Stephanie Slezycki

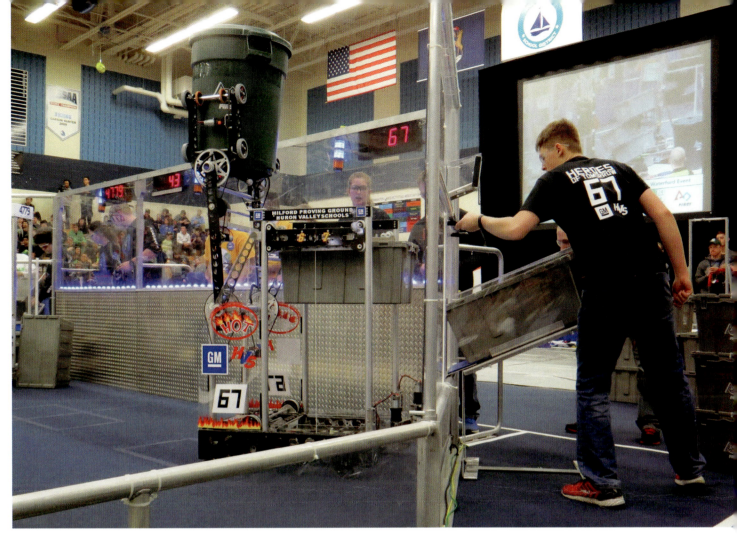

◐ Robots were designed to optimize the loading and stacking processes by minimizing the cycle time for loading and maximizing control of the collected cargo. Loading totes required a coordinated effort between the robot operators and team members who entered the individual totes into play.

The autonomous period was followed by a two minute and fifteen second session when human operators controlled the robots. The robot drivers would acquire totes from anywhere on the field, including the two loading stations in the corners of the field. Here team members would deposit totes on the field or in robots one tote at a time. Robotic elevators or arms would create stacks of the individual totes, with those stacks potentially topped off with a recycling container. The fact that the scoring platform was two inches tall and had a seven-inch-long, seventeen-degree incline from the playing field surface provided an additional challenge for stacking totes on its twenty-inch wide surface.

DESIGN, MANUFACTURING, AND CONTROL OF FRC ROBOTS

The process of creating a robot to compete in a *FIRST* event is a significant engineering challenge, amplified by the fact that the majority of the robot's design, manufacturing, and testing must be completed in six weeks. Design modifications are allowed during a short time period before each competition, thereby providing a mechanism for teams to refine and improve their designs during the competition season. FRC teams are free to organize themselves as they so desire, structuring their efforts in a way that best enables each team to motivate students to pursue careers in science, engineering, technology, and other professions. Teams typically consist of students who are mentored by teachers, parents, and professionals, with team sizes ranging between ten and one hundred members.

This book showcases the work of five exemplary FRC teams in each of six areas spanning the design,

◆ The raised scoring platform provided an additional challenge for the robot operators.

◆ In addition to the human-assisted loading stations, robots acquired totes from any location on the field including the close-packed landfill located at mid-field.

◆ Computer-aided design provided an efficient method to plan and execute ideas.

manufacturing, and control of robotic systems. The first collection of profiles illustrates how each FRC team skillfully applied computer-aided design, simulation, and analysis methodologies to establish, refine, and manufacture its robot. Noting that innovation is independent of technology, a collection of profiles highlights innovative design processes using traditional manufacturing methods. Modern manufacturing methods are explored in separate chapters covering 3D printing, as well as computer numerical control (CNC) machining and cutting systems. The final chapter addresses sensors, monitoring, and control applications implemented by FRC teams.

All of the information presented in this collection of team profiles was obtained by inviting 2,900 teams participating in the 2015 *FIRST* Robotics Competition to submit material for inclusion in this book. Leading examples of design, manufacturing, and control of robotic systems were selected from those submissions and are presented as design case studies in this book. Collectively these examples illustrate the value of a variety of design, manufacturing, and control methodologies used to create robotics systems.

◆ *FIRST* is a celebration of creativity, teamwork, and accomplishment, manifested by the design, creation, and control of sophisticated robots. These activities provide rewarding opportunities for team members to learn, develop, and grow.

THE POWER OF FIRST® MENTORS

FIRST® Mentors — Guiding Their FIRST Family

"In the inner city, boys are an endangered species." Such was the characterization of the low income neighborhood surrounding Crenshaw High School in South Central Los Angeles during the first decade of 2000, as described by long-time Crenshaw teacher Urban Reyes.

Urban began his career as a quality assurance engineer in the aerospace industry and later became a high school teacher when he was motivated to find a place where he could make a real difference in society. During Urban's tenure, Crenshaw High School faced numerous challenges, including having more than 80% of its families living below the federal poverty line and a population where the majority of students lived in foster care. Compounding these societal problems were a high turnover of administrators, and not surprisingly, below average performance in district-wide standardized tests. In 2013, Crenshaw High School was closed, to be reopened as a collection of magnet schools that provided an alternative form of education.

Even at its lowest point though, embedded in Crenshaw High School were people — teachers and volunteer mentors — who deeply cared about their students. In 2005, Urban founded FIRST® Robotics Competition (FRC®) Team 1692, the CougarBots, to help Crenshaw students see possibilities beyond their neighborhood. While Crenshaw's academic performance was the lowest of all of the 197 schools in the Los Angeles Unified School District, Urban's work became a beacon of hope in a challenging environment. Supported by companies such as the Aerospace Corporation and with the help of engineering mentors like Tim Wright — an Aerospace Corporation engineer — FRC Team 1692 developed into something much more than a high school robotics team. It became a clear path to a bright future.

Tim's motivation for being a FIRST mentor was similar to Urban's desire to become a teacher, because Tim wanted to make a difference in a meaningful way by sharing his knowledge of engineering and his own experiences with students. According to Urban, "Tim worked tirelessly at this — he gave it his all." Their experience with the students grew beyond leading an after school club. This team developed into a family.

The experience of two brothers — Yohance and Kumasi Salimu — provides insight into the role that FRC Team 1692 played in looking out for one another. Yohance, Kumasi, and their family lost their home in 2008, and spent the next few years staying with friends and living in a homeless shelter. This nomadic life provided many challenges to the brothers while attending Crenshaw High School. They were first drawn to the robotics team as an activity where they could spend more time at the high school, thereby being sheltered from the difficulties of life beyond the walls of Crenshaw High School.

Though Tim and Urban were initially unaware of the Salimu family's situation, they soon realized the complexities that Yohance and Kumasi were experiencing. As FIRST mentors who were committed their students, Tim and Urban made helping the brothers a priority and found various ways to offer assistance. The mentors made sure the brothers had safe places to stay, were fed, and were provided transportation to allow them to attend all of the robotics team's meetings. Tim's company provided the brothers with internships at Aerospace Corporation, with the brothers then using the income to help their family survive. The FIRST team quickly became pivotal in the lives of Yohance and Kumasi, and, according to Urban, put the brothers on the right side of the walk of life. Yohance and Kumasi thrived on their FIRST team, with Yohance developing as the team captain, later to be followed in that role by his younger brother, Kumasi.

The impact of FIRST did not end with graduation from high school for these two brothers, but rather their participation in FIRST propelled each to new opportunities. Yohance is now a cadet at the United States Air Force Academy and Kumasi is a student at Tuskegee University. Each is pursuing a science degree. Through the dedication and guidance of FIRST mentors Urban Reyes and Tim Wright, Yohance and Kumasi are on trajectories that otherwise would not have been even remotely possible.

FIRST mentors accomplish much more than simply helping students build robots. FIRST mentors guide students on the journey through life.

> FIRST is fueled by the dedication and commitment of mentors who guide students in technical disciplines, project management, and teamwork. Most significantly, FIRST mentors are role models who help students navigate the challenges of life. FIRST is much more than robots.

CHAPTER 2
💻 COMPUTER-AIDED DESIGN, SIMULATION, AND ANALYSIS

Computer-aided design (CAD) tools enable designers to transform creative thoughts and initial sketches into a mathematical description of those concepts in the digital domain. Creating a visual or digital model of what will ultimately be a physical object is the first step for all manufacturing methods. The digital model can be used to evaluate the design's utility, predict its physical properties, and manufacture the created part. In addition, CAD simulation programs can be used to import computer-generated models, investigate the effectiveness of the designed system, and evaluate the physics of the design when operated in the real world. Five profiles in this chapter illustrate common components of CAD software programs including sketching, three-dimensional (3D) model generation, animation, simulation, and performance analysis. The case studies emphasize the digital design and review process for systems that were created and evaluated using software developed by Autodesk, PTC, and SolidWorks.

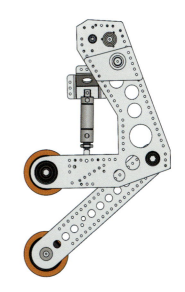

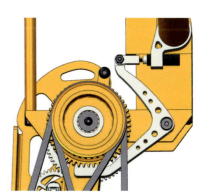

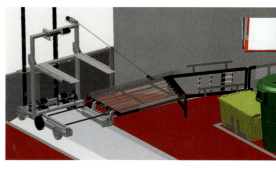

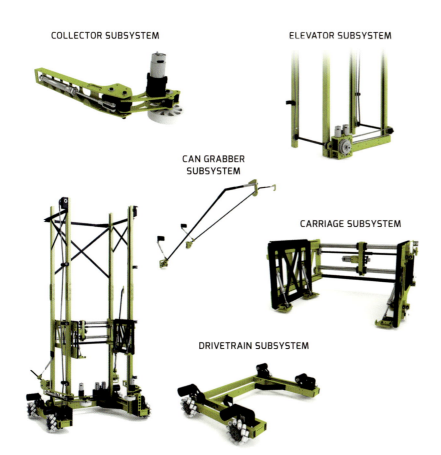

Computer-Aided Design, Simulation, and Analysis | 23

Team 118 – Champion of Design

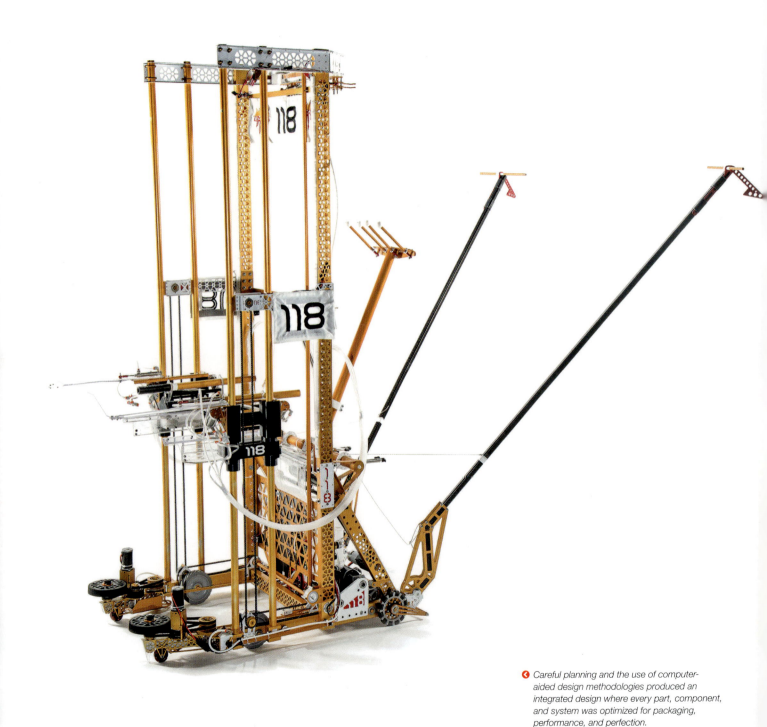

◀ Careful planning and the use of computer-aided design methodologies produced an integrated design where every part, component, and system was optimized for packaging, performance, and perfection.

ROBOT WILL – ROBOT DID

The path to become a champion of the *FIRST*® Robotics Competition (FRC®) for FRC Team 118, the Robonauts, began with a declarative plan written on a scrap of plywood. The team's "Robot Will" board detailed the robot's performance requirements following the team's close review of the 2015 FRC game challenge. Embedded in the list of 22 items were two that were especially important: drive through game pieces and move fast. This combination of robot skills demanded that the robot be capable of picking up totes and containers while moving forward, on the fly, and be able to do so at a high rate of speed. Equally important though was a design element added later: acquire containers from the center of the field during the autonomous period. This new performance criterion prompted the team's season-long quest to build the world's fastest "can burglar" — the FRC colloquialism for devices that acquired the center containers at the start of the match.

These three criteria, and the twenty other requirements prescribed on the "Robot Will" board, drove the team's design process. The goals also became a road map that was followed from the team's workshop at NASA's Johnson Space Center in Houston, TX, to the Einstein field at the *FIRST* Championship in St. Louis, MO. Progressing from the plan on plywood to the premier field at the *FIRST* Championship required detailed planning, careful construction, rigorous testing, and continuous improvement.

COOLEST ROBOTICS WORKSHOP ON THE PLANET AND BEYOND

Making the connection between a high school robotics program and the real world — and even out of this world — takes place every day at the Robonaut's robotics workshop. The team's expansive work area, which includes a complete playing field, is housed in a NASA astronaut training facility. As an open building with no interior walls, the work is in full view of everyone in the building, including over 700,000 annual visitors who observe activities from a twelve-foot-high balcony that spans one of the building's walls.

FRC Team 118's inanimate neighbors include an operational and full-sized Space Shuttle robot arm, the command

World-class design was developed in an "out of this world" workshop located at the Johnson Space Center in Houston, TX, where robotic systems for terrestrial and space applications were designed, developed, and tested.

After days of brainstorming and game analysis, the Robonauts' "Robot Will" board detailed the team's plan for success as performance criteria that guided the design process.

A nearby lunar rover and an original Robonaut provided inspiration and design concepts. These robotic systems examples illustrated the high levels of quality required for reliable operations and superior performance.

module for the new Orion Spacecraft, full-scale models of the International Space Station's (ISS) habitation modules, and a simulation section of the ISS solar panels. The team's neighbors also include international teams of astronauts, engineers, scientists, and technicians who are training for future space flights. The facility also supports current missions on the International Space Station by testing procedures before they are executed by the orbiting crew in space.

Other neighbors include concept exploration vehicles for manned missions to the moon, including one vehicle with a swerve drive system that was modeled after an FRC Team 118 robot drive system. The closest neighbor is a duplicate of NASA's Robonaut 2, the humanoid robot currently on board the ISS. The earth-bound version of Robonaut 2 watches over the team's practice field when not evaluating maneuvers to be executed by its ISS twin. A wall of FRC banners and former FRC Team 118 robots provide a backdrop and identify the mission of this section of the training facility. The inspiration within the facility has no boundaries, with the Robonauts team members inspired by their surroundings and NASA employees inspired by the work of FRC Team 118.

A HOME FOR WINNING SYSTEMS

The team's home within the astronaut training facility included areas for meeting, coding, computer-aided design (CAD), mechanical and electronics assembly, light fabrication, and, most importantly, testing. Machining and fabrication facilities were accessible in adjacent NASA machine shops. The full-sized field was created as soon as the game was announced and was always available for testing ideas, validating solutions, and identifying areas for improvement. These facilities provided the infrastructure that advanced the "Robot Will" board statements into real-world results.

Empire was completely modeled using PTC Creo software. Only after many hours of design, to the smallest level of detail and ensuring the full integration of all components, did the project advance to the manufacturing stage.

Detailed CAD plans were essential to convert pixels into reality for every piece on the team's robot. CAD was also used to redesign broken components and improve the performance of working components.

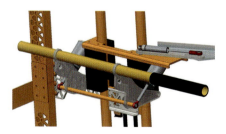

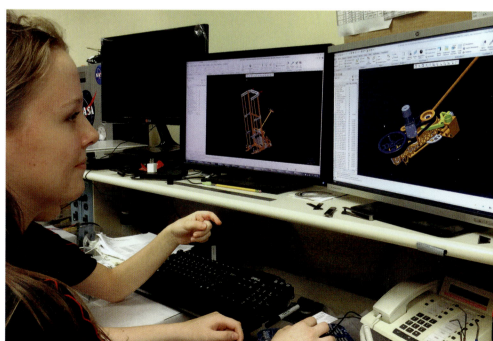

Computer-Aided Design, Simulation, and Analysis | 27

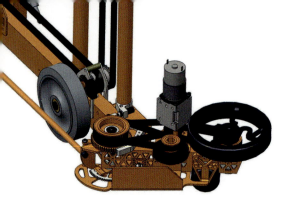
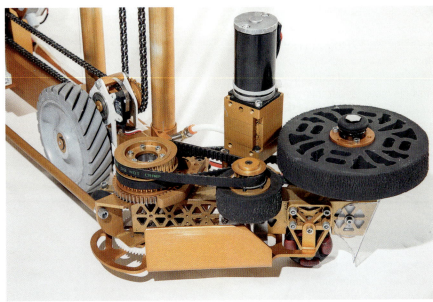

Pivoting the intake arms was accomplished with a unique power take-off mechanism. A miniature piston, linkages, belts, and gears allowed a single motor to power the intake wheels and open the intake arms.

Empire, FRC Team 118's robot, consisted of four major systems: the chassis, an intake system, an elevator, and a device to acquire containers from the center of the field. Another component was the container manipulator, with this system being less significant in terms of strategy and game play than the others. Each system was first designed by a sub-team using PTC Creo CAD software, with the CAD model constantly updated during manufacturing, assembly, and system testing.

The chassis and drive system were designed to be lightweight, simple, and reliable. By reducing the complexity in these components, more time was available during the design, manufacturing, and testing phases for the mechanisms that scored points. Stability and weight distribution were important factors in the chassis design. The four-wheel drive chassis had a minimal amount of ground clearance to maintain a low center of gravity. The drive motors and transmissions were located at the rear of the robot and mounted as low and as far back as possible to maximize their effects on stability.

Traction wheels located at the front of the robot and loaded by the weight of captured totes provided the robot's driving force. Non-powered omnidirectional wheels at the rear of the robot increased maneuverability. This arrangement allowed the robot to rotate around the center of a suspended stack of totes and eliminated an additional centrifugal force on the stack during rapid rotation. The omni wheels were manufactured by waterjet cutting 0.125-inch aluminum plate. The chain powering the forward wheel was routed through the two-inch by one-inch aluminum tube frame of the chassis. This arrangement constrained the chain within the tube, thereby reducing the chance for the chain to jump off sprockets, and protected the drive chain from entanglement with field elements. The tubes were cut on a manual mill using a digital readout to precisely locate the distances between bearing holes.

Prototyping and testing of initial designs for Empire's intake produced mechanisms capable of ingesting totes at a rate of one tote per second. This was accomplished by approaching target totes at nearly any angle, satisfying one of the guiding "Robot Will" principles. The intake system consisted of a number of components including gripping wheels, motors, articulating arms, anti-tipping wheels, and a sensor to detect when a tote was engaged. After evaluating many options, the gripping wheels at the mouth of the intake were manufactured by waterjet cutting one-inch-thick, high-strength neoprene rubber sheet. The tackiness of the wheels was ensured during each match by cleaning the wheels with acetone.

Neoprene rubber bands on the arms acted as a spring to squeeze the captured totes with approximately ten pounds of force. While this passive resistance pulled the arms in, the springs also allowed the arms to open as wide as needed to ingest totes. A single motor and transmission powered three devices on each intake: the gripping wheel, a guide wheel, and a drive mechanism to open the intake for scoring the totes on the platform. This efficient use of one motor for three functions was produced using a series of belts, appropriately sized pulleys, and a clever pneumatically actuated mechanism.

A rack and pinion mechanism opened the arms to release stacks of totes. To open the arms, a small pneumatic piston energized and pulled the rack into the motor-driven gear. Since the gear was also the arm's pivot point, the rotating motor pushed the engaged rack, which, in turn, opened the arms. To close the arms and disengage the rack, the piston retracted to disengage the rack from the gear, with the arms then closed by the force of the neoprene springs. The rack was designed to dig into the gear when the motor ran in the direction to ingest totes, and to bounce away from the gear when the motor ran in the opposite direction.

The current powering the intake motors was monitored to prevent the motors from stalling due to misaligned totes. Proximity sensors determined when a tote was fully engaged. Automatic control sequences commanded the rotation of the intake wheel to ingest the totes, including momentary reversal to release the stuck totes, and coordinated the piston and intake motor actions to open the arms and release the stacks of totes.

The technique of linking multiple force delivery systems to a single motor was

A powered elevator lifted totes and containers while a pneumatic gripping system that increased the stack's stability was passively lifted by the tallest object in stack.

Computer-Aided Design, Simulation, and Analysis | 29

After eighteen years of participating in the FIRST Robotics Competition, the Robonauts completed a banner season of successful CAD design and careful manufacturing as the captain of the FIRST Championship winning alliance.

also applied in Empire's elevator and stack restraining system. Four circular aluminum tubes served as elevator guides for a pair of carriages on each side of the robot to lift and restrain stacks of totes and containers. The carriages were fabricated on three-dimensional (3D) printers, allowing for each carriage to be optimally designed for its specific purpose. The carriages lasted all season, which was testimony to the integrity of the 3D-printed parts.

The lower carriage was fabricated on a Fortus 400ms 3D printer using ULTEM, a thermoplastic with exceptional mechanical, thermal, and chemical properties. The ULTEM's high wear resistance allowed the carriage to ride on the rails as its own bearing surface. The ULTEM's material strength enabled the carriage to serve as a foundation for mounting the components needed to make the device operational. The combination

of a low friction bearing and circular shafts eliminated binding between the carriage and the supports. Passive Lexan fingers attached to the carriage to grip the totes and containers.

Two motors and a transmission mounted at the rear of the robot powered the elevator, with an aluminum shaft transferring power across the robot's frame. A horizontal loop of chain connected to a vertical loop of chain that rotated around a sprocket mounted between the elevator shafts. A string potentiometer housed in a custom fixture measured the elevator height.

In addition to the lifting carriages, a pair of restraining carriages also passively rode on the elevator shafts. The restraining carriages, which were 3D-printed using acrylonitrile butadiene styrene (ABS) plastic, were lifted by the totes and the container. Attachments on these carriages provided side forces to each game

piece to keep the load contained inside the elevator. Pneumatically controlled linkages tightly held either the highest tote or the container to lock the top of the stack to the robot's elevator.

The restraining mechanism was automatically released when the stack was scored, with the upper carriage then falling back on top of the lifting carriage to repeat the stacking cycle. This system allowed Empire to travel quickly around the field while securely holding a stack of totes capped by a container, with this ability greatly contributing to its on-field success. As with the use of a single power source for multiple operations in the intake system, here, too, a single power source provided many operational functions for the elevator system.

The mechanism to acquire the center platform containers was named "Can Domination," a reflection of the team's prediction for its performance during this component of the game. This system evolved over the season, from continuous refinement that decreased its Championship deployment period to a small fraction of its original grabbing time of 330 milliseconds. The first generation of the device used in the team's regional competitions was a multi-component system that was manufactured using a lathe, waterjet cutter, tube bender, and 3D printer.

While the device was effective, the Robonauts wanted a better system and devoted seven weeks during the competition season to create the fastest "can burglar" in FIRST. Goals of the redesign included minimizing the mass of the device to maximize its deployment speed, and optimizing the use of springs as the mechanism's actuator. Stiffness of the deployed arm was essential to minimize stress and impart all of the stored energy into the deployment. This combination of factors led to the use of a spring-loaded carbon shaft, armed with an aluminum barb and secured to the frame in a pivoting sheet metal base.

Using sheet metal for the system's base allowed for a 24-hour iteration period at the team's workshop to produce the final component. Eight pieces of metal of different thicknesses provided the needed strength and fit together like a puzzle. The pieces clamped the carbon fiber tube and preserved the tube's structural integrity since the carbon fiber was not perforated with attachment holes. Strands of neoprene tubing were attached to the bottom of the base below its pivot point and provided the force to activate the system. Five hundred inch-pounds of torque powered each arm and were responsible for the system's high rate of speed.

The tubing was loaded when the arms were stored within the chassis footprint. Small pneumatic pistons armed a mechanism to hold the arms in place. At the start of the match, the pistons extended, the restraining mechanism was released, and the arms slammed into the

◐ Power for the elevator lift was transferred using vertical and horizontal chains, as well as a cross-robot jack shaft.

◐ Numerous design iterations of the container-grabbing mechanism over the competition season shaved precious milliseconds off this system's initial cycle time.

Coordinated actions simultaneously released the upper gripping arms, lowered the stack, opened the intake arms, and backed up the robot to deposit stacks on the scoring platform.

containers. The barbs that snagged the containers were attached to the carbon fiber tubes with gaffer tape to minimize weight and preserve the structural integrity in the tube section having the highest rotational velocity. Vectran string connected the carbon fiber arms to the robot frame and provided a hard stop to minimize the slamming load on the arms. A winch pulled the arms back into the robot's perimeter following their deployment.

A period of continuous redesign and refinement perfected this system. As one example of the level of detail devoted to this system, the final barbs to snag the containers were optimized after sixteen iterations of the design. The final speed of the system was 180 milliseconds, validating the team's energy expended on this system.

A final device on the robot was one to acquire containers after the start of the match. The team called this device its pike, which was constructed from the installed remnants of a previously designed but unused arm. After backing the robot toward the center of the field, this arm grabbed the container's handle as the robot dragged the container onto its alliance's side of the field for later use.

OUT OF THIS WORLD SUCCESS

Empire's hardware solutions were partnered with an equally effective and reliable software control system. The hardware and software systems were fine-tuned throughout the season using an iterative process to optimize speed and performance. CAD software was an important design tool at each step of the process and especially so during iterations to ensure that the actions of all systems were integrated with each other. The CAD software also produced the necessary plans to manufacture and remanufacture components using lathes, computerized numerical control mills, waterjet cutters, and 3D printers.

In the end, Empire met the design goals recorded on the "Robot Will" board and more. That success led to the team winning three regional competitions and the *FIRST* Championship. The year's work produced four new champion banners and a great addition to the Robonauts' Robot Hall of Fame that will inspire future astronauts, engineers, technicians, scientists, and visitors at NASA's Johnson Space Center.

The intake was rigorously tested to determine the optimal tension force to ingest totes. Optimization of this subsystem and other components produced an intake and elevator system capable of building a six-stack of totes in ten seconds.

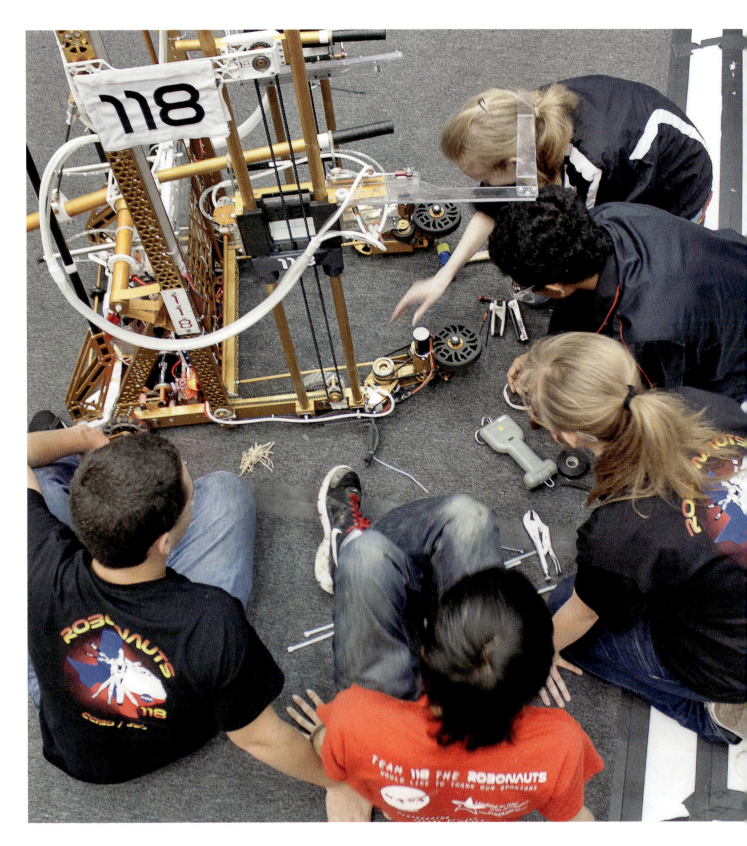

Computer-Aided Design, Simulation, and Analysis | 33

Team 846 – Monkeying Around with Gears

GEAR COMBINATIONS THAT WORK

Computer-aided design (CAD) was fundamental to the success of FIRST® Robotics Competition (FRC®) Team 846, the Funky Monkeys, of Palo Alto, CA. A robot with five independent modules was produced by five student-led teams that designed each module. Each team maintained that module's design files using version control software to share files and protect the integrity of design changes. The software allowed all files to be maintained in a central database and provided access to the files to all students. This approach synchronized design changes among the different groups no matter when the sub-teams were working. The version control software also streamlined the process to convert concepts into working mechanisms and integrate individual components into a functional robot.

The use and manipulation of gears to transmit torque from a power source to its application was common to each design sub-team. In one configuration, a transmission was designed to easily change gears and obtain different performance characteristics. In another application, gears were arranged to position belt-driven motors directly above the wheels they were driving, thereby maximizing the effect of the motors' weight over the drive wheels. In yet another application, gears amplified the rotary motion provided by a linear piston. The Funky Monkeys demonstrated a keen ability to monkey around with gears to achieve the best performance.

A CLEVER CARRIAGE

The carriage that lifted the totes and the containers was cleverly designed to grasp these very different objects in any orientation. The carriage consisted of four hooks to grab either object. The profiles of each object were analyzed with Autodesk Inventor CAD software to determine the shape, size, and spacing of each hook. The hooks were designed to conform to the shape of the totes and containers and had specific features for manipulating each object.

Included in the carriage was a motor-driven system that extended the gripping mechanism outside the robot's perimeter. The extension enabled the robot to place a stack of totes on the scoring platform without having to drive on the platform, thereby ensuring that the stack remained level and stable. This sliding mechanism was driven by a motor and deposited the cargo with precise control. The system to grasp and deposit both totes and containers was designed to maximize performance using a minimum number of manufactured parts.

FLEXIBILITY FROM WORM GEARS AND SPRINGS

The carriage was lifted using a motor-driven system augmented by a counterbalance mechanism, similar to systems used in commercial elevators. Two motors and a transmission provided lifting power and used a chain to link the lift mechanisms at the front and back of the elevator. Solid shafts connected each side

Five sub-teams developed five modules that, when combined, created a winning robot. Version control software ensured the coordination of the distributed CAD process.

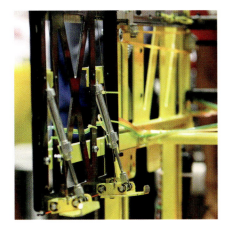

The lips of the pneumatically activated hooks keyed into the edges of the totes, while the hooks' curved edges captured the containers.

A mecanum drive system provided the speed and maneuverability for quickly aligning the robot to acquire, transport, and deliver totes.

of the lifting system and transmitted torque from one side of the robot to the other. These linked systems lifted each corner of the carriage in unison, a technique that provided smooth lifting and prevented binding.

A worm gear transmission connected to the two drive motors was designed to operate the lift at two different speeds. In the fastest lifting speed configuration, large gears were coupled to the motor drive shafts using a smaller spur gear. For greater lifting capacity, the size of the drive shaft gears was reduced and the spur gear size was increased. Since the sum of the teeth in the gear train remained constant, the center-to-center distances between the gears was also constant, and the two different gear combinations could be inserted in the system. The final design lifted six totes at a rate of two feet per second.

To reduce the load on the motors and increase the system's operating speed, a rotary spring was added as a counterbalance. A Neg'ator constant torque spring motor was mounted inline with a spool attached to the powered drive shaft in the transmission. A cord connected the counterbalance spring to a spool and the size of the spool was altered to change the magnitude of the torque delivered to the transmission. The simplicity of this device and its ability to easily adjust the delivered torque, combined with its compact profile, produced a reliable and effective mechanism.

BELTED DRIVE

The team's mecanum drive system was designed to create a fast and nimble power train. While this drive system had little resistance to external forces, such as encounters with another robot, it allowed the robot to move in any direction and any orientation. This mobility was an essential feature to easily position the robot in tight quarters and optimize its ability to acquire totes and containers. Typical installations for mecanum drive systems have the drive motor in the same plane as the wheels, resulting in a configuration that consumes a large portion of internal volume inside the robot base.

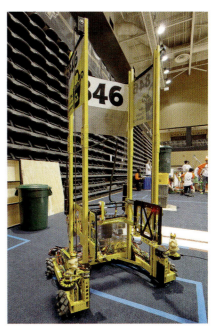

The integration of the drive, intake, grasping, carriage, and elevator systems relied on CAD planning to ensure all systems fit within the assigned volume.

◉ A counterbalanced elevator system was lifted on each side by chains and powered by two motors from a single gearbox. The elevator could lift six totes and a container at a speed of two feet per second.

◉ The elevator carriage was a multi-purpose mechanism capable of grasping totes and containers in any orientation. A belt drive extended the carriage on slides to deposit stacks on the scoring platform.

◉ Additional lifting torque was provided by a spring attached to a spool on the right of the gearbox.

A belt drive was used to provide an alternate orientation — mounting the motors above the wheels. This arrangement preserved valuable room at the base of the robot to ingest totes and containers. The belt system also served as the first stage of the drive transmission. Gears provided an additional speed reduction from the motors to the wheels, with these parts held between plates that also supported the motors and wheels. The compact transmission was mounted between box beams that comprised the frame and provided a solid foundation for the drive system.

GEARING MORE SWING

Another ingenious use of gears was applied in the collector system where powered wheels gathered totes and containers. The collector arms were stored within the robot perimeter and swung outside the robot frame to gather objects. Pneumatic cylinders provided the force needed to pivot the collector arms into position.

The required piston force needed to move the arms and the geometry of the compact design that packaged the actuator and the collector arm posed a design challenge. The arms needed to move through an angle of 160 degrees between their stored and ready conditions to effectively collect totes and containers. The optimum piston that could supply the needed force for this application and simultaneously fit into the designated space on the robot could only provide 120 degrees of rotation, 40 degrees short of the required angle.

To solve this problem, the cylinder clevis was coupled to an arm attached to a 40-tooth gear. This gear drove a 30-tooth gear that attached to the collector arm. Since only one third of the 40-tooth gear was needed to obtain the required 120 degrees of rotation, the gear was modified by cutting away one side of its face. This allowed the cylinder's clevis to be mounted in the same plane as the gear, thereby resulting in a very compact system. The shape of the connecting plate between the clevis and the 40-tooth gear was designed using Autodesk Inventor's interference analysis and motion constraint tools to identify material overlaps and produce a compact design.

◉ Linked mechanisms powered by a single source ensured that both sides of the elevator carriage moved at the same speed and minimized binding.

◉ Long carbon fiber arms were actuated using pneumatics. A pulley system extended internal aluminum tubes from the upper tubes to reach the center containers.

The last stage of optimizing the intake involved finding the best wheel surface to ingest totes. The load placed on the totes caused by pinching them between the intake rollers, as well as the load due to traction of the wheels on the totes, was analyzed to understand the effect of friction between the rollers and the totes. This force analysis was followed by testing, using a variety of wheel configurations to determine that rubber-treaded wheels were optimum for this application.

QUICKLY GRABBING CONTAINERS

The team's "can burglar" — the device to grab the center containers quicker than the opposing alliance — consisted of carbon fiber arms and an aluminum tube. The aluminum tube was contained inside the carbon fiber arm and telescoped out at the start of the match. Pneumatic actuators powered the arms to grab the containers in the center of the field.

The inner tubes were pulled out from their stored position by a line attached at the end of each arm. As the pneumatic cylinder actuated the arm into position, the fixed string

◉ Pneumatic cylinders pivoted each of the collector arms from inside the robot's perimeter into the optimal position to collect totes.

◉ A hybrid belt and gearbox transmission positioned the drive motors over the wheels to maximize the available area to collect totes.

Computer-Aided Design, Simulation, and Analysis | 37

FRC Team 846, the Funky Monkeys, from Lynbrook High School in San Jose, CA, is a student-run, extracurricular organization with over 100 members. In 2015, CAD was used as a tool to simultaneously engage a large number of students in the design process.

The use of CAD was essential to combine multiple systems, including power sources and actuators, into a small volume.

extended the inner extensions. Three-dimensional-printed hooks at the end of each arm latched onto the center of the container and pulled the containers into the landfill as the robot drove toward the player station.

1,500 CHANGES FOR ONE SUCCESS

The modular design and version control software provided a system to review each mechanism independently. During the season, over 1,500 changes to the robot design were cataloged and stored in the file repository. This system allowed the sub-teams to work autonomously, with all changes automatically shared with the entire team.

The robot was a top performer, consistently capable of making stacks of six totes topped with a recycling container. Augmenting the robot's performance was its appearance. The use of CAD as a planning and manufacturing tool resulted in a stylish and high-scoring robot. The Funky Monkeys may have monkeyed around a bit with gears, but their results were far from being labeled as funky.

❯ *A coordinated use of CAD allowed five independently designed subsystems by five groups of students to be integrated into a high-performing robot. Version control software was used to share files and synchronize design alterations.*

Team 1114 – Computer-Aided Design as One Step of the Design Process

> This rendered CAD model illustrates all components of the robot, including structural components, mechanical systems, pneumatic actuators, and motors, as well as the robot's power distribution and control systems.

EXECUTION OF A PLAN ENSURES SUCCESS

Success in the *FIRST*® Robotics Competition (FRC®) does not come by chance. Success, both on and off the playing field, is the result of careful planning and the execution of many details. While there is no guarantee for strong performance, the application of a rigorous design process — from concept to completion — increases the odds for success. The integration of teamwork and the careful application of resources, including the use of tools to guide the design process, are also keys for success.

FRC Team 1114 has over one decade of outstanding performance designing and building super competitive robots. To do so in a short period of time with a large and distributed team requires leadership and decision making coupled with an effective use of design and manufacturing tools. For this team, the use of computer-aided design (CAD) allowed the team to maximize its effectiveness from the original concept to the final moments of the competition.

THE DESIGN PROCESS

FRC Team 1114 applied four distinct phases to create its robot and compete in the *FIRST* Robotics Competition: conceptualizing a solution; component design, prototyping, and integration; manufacturing and assembly; and testing, revising, and competing. This process spanned the entire competition season from the announcement of the annual challenge to the final match. As an iterative process, revisions were

⬤ The form, function, and actions to lift a stack of totes and ingest individual totes below the lifted stack were coordinated to minimize the time required to load totes into the robot. The integration was planned in the initial design stage using CAD software.

made as late as the final match to apply all of the season's accumulated knowledge and experience.

The first phase of the FRC Team 1114 design process was the development of strategy to play the game, the creation of objectives that achieved this strategy, and the initial design of robot systems and components that fulfilled the objectives and executed the strategy. The entire team worked as a group to develop the strategy after carefully reviewing the game rules. In 2015, that strategy included maximizing scoring by being maneuverable and efficiently manipulating totes and containers. That general strategy established a set of specific objectives for the robot to accomplish. Brainstorming and other idea generation techniques were used to generate concepts that achieved the objectives. This phase was completed once the robot systems and components were identified.

The team's objectives for the 2015 season included a highly agile robot drive system, an ability to acquire totes from the landfill, the capability to stack totes on the robot, manipulating recycling containers in any orientation, creating stacks of totes and containers, depositing stacks on scoring platform, and acquiring containers from the center of the competition field. Based on this set of defined objectives, the robot systems that were needed to achieve these tasks were conceptualized by groups of team members with each group designing a specific system.

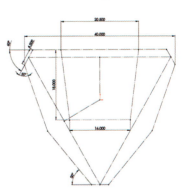

⬤ Two-dimensional analysis was used to establish the robot's perimeter and the footprint for key components, with these sketches later developed into three-dimensional models.

Computer-Aided Design, Simulation, and Analysis | 41

◯ *Intake wheels spinning at 800 RPM gripped the totes and quickly pulled them into the robot. The outer wheels were spring-loaded and conformed to the orientation of the totes.*

The next phase of the design process developed concepts that converted the established design features into working systems. The components for each robot system were designed in this phase, with the design progressing from initial sketches to detailed plans that included all parts as a computer model of that system. Ideas were quickly tested as prototypes to validate the design approach for each system. Integration of the systems was also included in this phase of the design process, with the separate systems combined using CAD tools to create the composite robot. This model had a high degree of fidelity and included nearly all of the robot's components, including most of the electrical system.

Once the robot was defined as a CAD model, the individual parts were manufactured and assembled. For FRC Team 1114, the manufacturing process entailed exporting the CAD files for each structural member to the team's sponsor — Innovation First International, located in Texas. Once the aluminum sheet pieces were manufactured, they were shipped to the team in St. Catherines, Ontario. The structure was assembled using rivets and the mechanical and electrical

◯ *The elevator was tilted back three degrees to support stacks of totes. The robot base rotated forward three degrees when scoring totes to perfectly align stacks of totes on the scoring platform.*

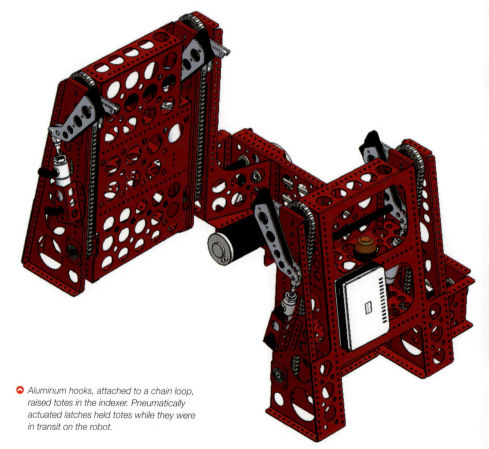

◯ *Aluminum hooks, attached to a chain loop, raised totes in the indexer. Pneumatically actuated latches held totes while they were in transit on the robot.*

42 | *FIRST* Robots: Behind the Design | Vince Wilczynski and Stephanie Slezycki

◆ A single arm with two extensions grabbed the containers from the center of the field. Surgical tubing pulled the mechanism into position at the start of the match.

components were then added to the assembled frame. Because the robot was modularly designed, the assembly of each module was independent of the other modules and allowed many people to work simultaneously. The integration of the individual systems into a functional robot completed this phase of the design process.

The final phase of the design process was a continual period of testing, revising, and competing. Since the entire robot existed as an accurate CAD model, it was easy to alter the design and create new components for improving performance. To increase its competitiveness, the team constructed two robots, with the auxiliary practice robot being an exact copy of the robot used in actual competitions. By having both an accurate CAD model and a duplicate robot to test ideas after the build season was completed, the team used these resources to rigorously test and refine ideas that would increase the robot's performance.

DESIGNING KEY COMPONENTS

From the earliest stages of the design process, CAD was used as a tool to accelerate creativity. Robot components were first sketched using SolidWorks CAD software to identify the critical dimensions and integrate systems. Prototyping of individual systems established these critical dimensions and the ultimate location of motors, manipulators, and power transmission components. The design process advanced based on these sketches and prototype findings, with accurate CAD models then created for each independent system. Communication between designers during this process ensured that the independent systems could be integrated as a complete robot.

A three-wheeled holonomic drive system provided a high degree of maneuverability for quickly positioning the robot when picking up totes. With a holonomic drive the robot could instantly move in any direction and in any orientation as if it was floating across the field. Without a need to worry about defense, the low traction of this drive system was not an issue.

The drive consisted of three omnidirectional wheels, each separately powered, located at the vertices of an equilateral triangle. A unique aspect of this design was the overall footprint of the robot base. Given the necessity to ingest a tote, the geometry established in the original CAD sketch mandated a base that exceeded the robot transportation rules in its playing orientation. To meet this rule the robot base had to be transported on its side. As such, all

CAD software was used to create the team's concept, prototype, and detailed design plans.

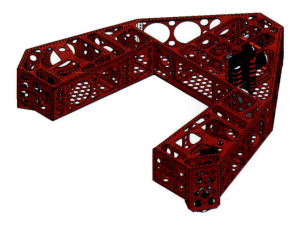

A three-wheeled, six-motor holonomic drive system using omnidirectional wheels produced a highly maneuverable robot. This drivetrain propelled the robot at a maximum speed of 11 feet per second.

The base geometry dictated that the robot be transported on its side. This transportation orientation required that appendages over 28 inches in height be attached to the robot on the competition field.

vertical components extending beyond 28 inches from the bottom of the robot were detachable and installed on the robot when placed on the playing field.

The intake consisted of a pair of wheels driven from a common motor spinning at 800 revolutions per minute. The outer wheels were spring-loaded to conform to the orientation of the totes in the landfill, a feature that allowed the robot to approach totes at nearly a 90-degree angle. Pneumatic actuators manipulated the intake to acquire containers and to release a stack of totes on the scoring platform.

Totes were lifted using aluminum hooks attached to a continuous chain loop. The hooks grabbed the edges of the totes to lift them from the floor, with pneumatically actuated latches used to hold and release the suspended totes. This system, known as the indexer, relied on a series of sensors to coordinate the mechanisms used to lift, hold, and release stacks of totes. Also included in this system was a pneumatic cylinder to push a stack of totes onto the raised platform in the center of the field.

The claw and vertical supports were an integrated system to grab recycling containers and stabilize the growing stack of totes below the container. The supports consisted of one-inch by one-inch tubes angled back at three degrees. This geometry helped support the totes while the robot moved around the competition field. In addition, because the robot base tilted up three degrees while dropping a stack of totes on the scoring platform, the three-degree offset ensured that each stack was exactly vertical when deposited on the scoring platform. The claw was attached to a passive carriage that moved up and down on the vertical supports, with the indexer providing the power to lift the totes and container.

A boot mechanism was designed to rotate horizontal containers using the bottom of the container as the mechanism's pivot point. Once the boot was pneumatically actuated into a deployed position, the intake wheels provided the power to ingest the container with the mechanism then righting the container as it was ingested and secured with the claw. To achieve the design

> Using a pivot point at the bottom of a container, a boot mechanism rotated containers to ease loading and stacking. The intake wheels provided the power to ingest and rotate the containers.

objective to acquire containers from the center of the field, a mechanism was designed to deliver hooks into two containers simultaneously. Preloaded surgical tubing deployed the hooks, with the robot then driven forward to pull the hooked containers onto its side of the field.

CAD ADDS VALUE

Computer-aided design was an important aspect of FRC Team 1114's success. This design tool allowed many different systems to be tested and iterated upon, with the CAD models and prototypes developed simultaneously. This established an efficient process to quickly modify, test, and evaluate ideas. Later in the design process, the CAD model was used to refine the design and incorporate additional improvements. As a digital model, it was much easier to edit and evaluate a digital component compared to machining a physical part to test an idea. Adding new components to the robot was also an easy process, as those components could be added virtually to the computer model. Once the new component's function and fit was verified in the CAD model it could be manufactured and installed with a high degree of certainty for its performance.

CAD tools were also used to determine the weight of system components, with material subsequently removed in the early stages of the design process to optimize the weight of each component while preserving functional performance. Finite element analysis was used to ensure that the designed structures were sufficient to support the expected loads in each member.

The CAD files were stored on the cloud-based platform Dropbox, with the team using this remote server to archive design revisions, parts lists, and layout drawings. Using remotely stored files, students and mentors could work side-by-side even when they were not physically located in the same place. Access to the files promoted collaboration among team members who were working on different components of the robot. The remote storage of the CAD files also facilitated the fabrication process used by FRC Team 1114 when components were manufactured one thousand miles from the team's home city.

Defining the robot in CAD was also helpful for manufacturing and assembly because the plans increased the student team members' abilities to perform these tasks independently. CAD drawings were provided to students as instruction sets for fabricating structural members, manufacturing parts, assembling systems, and installing electronics. The CAD plans were then used to review the independent work and verify accuracy.

PHYSICAL AND VIRTUAL DESIGN LEADS TO SUCCESS

Designing and constructing a sophisticated robot in six weeks is a significant challenge. Designing simultaneously in physical and virtual domains has the potential to increase a team's efficiency and effectiveness. The performance of FRC Team 1114 illustrates the value of applying a structured design process, relying on CAD for conceptual and detailed design, and iterating to make improvements.

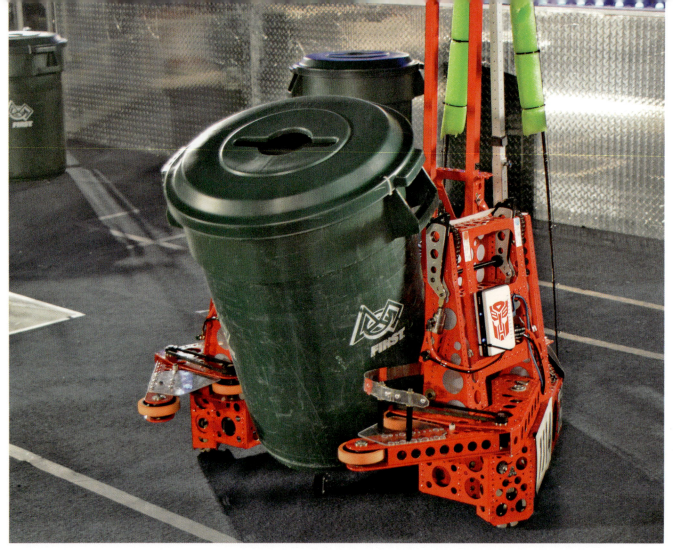

🔺 All elements of the robot's base, intake, indexer, and elevator were designed as a highly integrated system to acquire, stack, secure, and deliver totes and containers.

For this team, the CAD model and the physical robot are, in one sense, the same entity. An initial design was created to achieve the team's strategy objectives, with team members designing separate systems to fulfill specific objectives. These independent systems were then combined to produce a composite robot design. Working first in the CAD realm allowed team members to visualize the entire robot before it was built.

Seeing the robot in its totality before it was constructed greatly benefitted the team's modular approach to design and enabled systems to be simultaneously designed, manufactured, and constructed. These time saving applications gave FRC Team 1114 the opportunity to leverage its abilities and devote resources to activities that had the highest payback. By doing so, the team's effectiveness and chance for success on the competition field was greatly increased, which was evidenced by its success while participating in the *FIRST* Robotics Competition.

▸ A pneumatically actuated claw provided 40 pounds of container-gripping force. The claw was passively mounted on the elevator uprights and relied on the indexer to be lifted. The system optimized gravity, minimized power needs, and stabilized stacks.

Team 1511 – Many Uses of Computer-Aided Design

Autodesk Inventor CAD software was used to design the robot, analyze forces, calculate load safety factors, estimate the robot's center of gravity, and produce fabrication drawings.

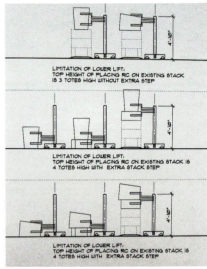

The design process began with a two-dimensional analysis to examine the robot's potential geometry and establish the design's criteria and constraints.

DESIGNING IN STAGES

Often designs progress from a concept to a sketch, followed by refinement of that sketch to more detailed plans that allow the device to be manufactured and assembled. This process can be made more efficient with the use of computer-aided design (CAD) tools. Degrees of sophistication can be added as the product moves along the design path, including using advanced features in CAD software to estimate parameters and predict performance.

This was the creative route followed by *FIRST*® Robotics Competition (FRC®) Team 1511, Rolling Thunder, from Penfield, NY, for the design of Thunderous Rex, a name inspired by

Exact CAD models of each of the robot's systems served as a means to evaluate mechanisms, predict system loads, fabricate parts, and guide the assembly of individual pieces into functioning components.

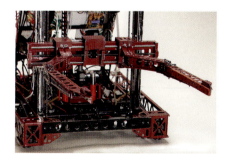

the robot's profile. The team started the design process with initial concepts that were converted to two-dimensional (2D) CAD models and later refined to become three-dimensional (3D) models of each component. Once the model had sufficient definition, CAD software features such as animation, weight prediction, and deformation analysis were applied to learn valuable information about the robot.

COMPUTER-ASSISTED DESIGN PROCESS

Hand sketches of potential robot configurations established the team's foray into creating a robot. These ideas were refined as 2D CAD drawings using Autodesk AutoCAD software to review the viability of the proposed design with regards to field element geometry and robot performance requirements.

For example, a preliminary profile view of the robot determined the height required for the lifting mechanism. The initial sketches advanced into more detailed two-dimensional sketches in plan and profile orientations that provided clarity on the design. Two-dimensional sketching as a design tool was not only used to define the robot's outline, but also used to evaluate specific mechanisms, such as alternatives for changing the spacing between arms for grabbing containers and totes. This first stage of work produced estimates for the locations of

Computer-Aided Design, Simulation, and Analysis | 49

all major systems and final dimensions of key components. These dimensions included the range of motion for the gripping arm, the height of the elevator, and the placement of the wheels.

At this point the robot was established conceptually as a collection of three fundamental systems that needed to be designed: a drive base, elevator, and gripper. CAD modeling progressed to three dimensions using Autodesk Inventor software. Once individual components were created they were combined to create a complete model of the robot. The creation of detailed component models allowed the use of advanced CAD features to further evaluate the design.

Additional CAD tools added to the clarity of the design. Finite element analysis predicted that the deflection of the gripping arms was not significant, indicating no need to alter their structural design. Animation was used to verify that the range of motion for the gripping arms met the design requirement established in the 2D analysis. Once the material properties of the robot were established in the CAD model the weight of the robot was estimated. These properties were also used to predict the location of the robot's center of gravity as being 17 inches high and 23 inches from the front of the robot. This prediction of the location of the center of gravity aligned with an earlier estimate calculated by hand (10 inches high and 23 inches from the front of the robot), thereby adding credibility to the independent estimates.

Digital organization was essential to maintain the team's large number of part files and assembly drawings. Every part was assigned a unique part number, with this part number appearing in every computer file name related to that object. Similarly, every assembly had a unique assembly number. File storage was based on an organizational structure for each component, with the parts and assemblies used in that component maintained in separate folders. The part numbers were also logged in

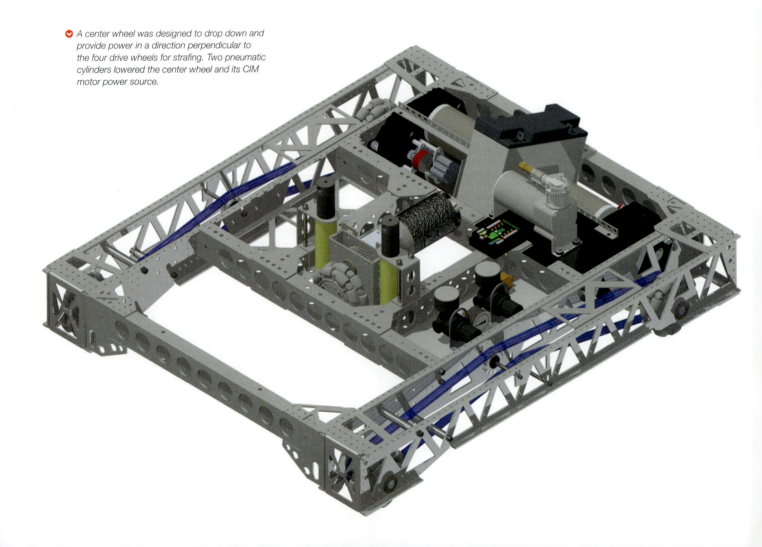

◆ A center wheel was designed to drop down and provide power in a direction perpendicular to the four drive wheels for strafing. Two pneumatic cylinders lowered the center wheel and its CIM motor power source.

a remotely accessible spreadsheet to keep track of the revisions and to record data such as the work date, level of completion, and name of the person working on the part. This log was also used to monitor the fabrication of each part, including the generation of a fabrication drawing, location of fabrication, and delivery date of the completed part.

Mimicking this approach, an assembly log was created to follow the progress of the components and their integration into the final CAD model of the robot. To minimize errors such as accidental file overwrites, incorrect updates, and misplacement of files and components, only three students were permitted to work on the final robot assembly. Given the number of individuals working on the project and the volume of parts, keeping track of all elements of the process was mandatory.

DRAWING TO ASSEMBLY

The design, fabrication, and assembly process of the gripping system illustrates how an effective methodology was used to produce components. The early sketches of this mechanism called for a gripper and carriage that would be lifted by the elevator. Detailed design began with the creation of the 3D model of this system. Where possible, existing part models for purchased components — such as bearings, motors, and belts — were imported and combined with team-generated parts to create a 3D model of each component.

Autodesk Inventor was used to create the fabrication drawings of the modeled parts, as well as the necessary files for manufacturing components using waterjet cutters and computer numerical control (CNC) mills and lathes. The automated work was completed by the team's sponsors: Harris RF Communications and Chamtek Machining. Parts were also procured and manufactured by the students in the school's workshop. The completed parts were identified using the standardized numbering system used in the CAD models. The accuracy of the completed parts was verified using the CAD model, with the parts then assembled to create the robot.

Assembly drawings served as a guide for combining the completed parts into working systems. Screen captures of the CAD components were used

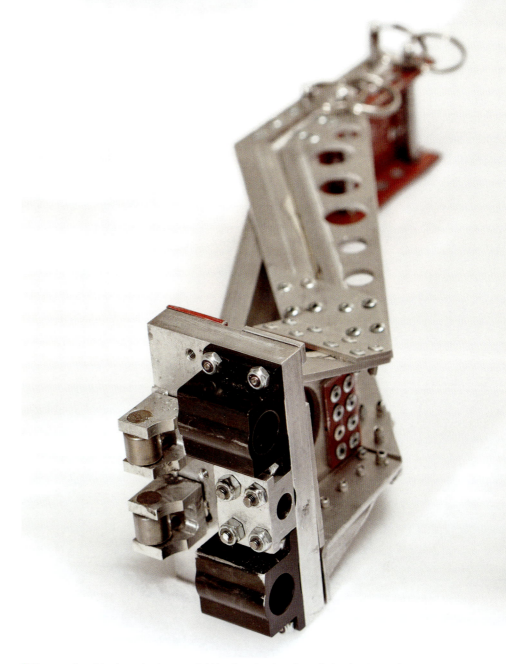

The mounting plates for each gripper arm held two linear bearings, two roller bearings to transfer lateral loads to the carriage, and a brass nut that was powered by a lead screw and moved each arm to grab totes.

CAD was needed to integrate systems within the small footprint of the robot's base. CAD plans were used by in-house and external fabricators to create the parts that were assembled into a robot.

to establish the assembly sequence. Completed components were then installed on the robot and tested.

FITTING IN A DRIVE SYSTEM

The drive system consisted of two four-inch omnidirectional wheels on each side of the robot, with the pair of wheels jointly driven by two CIM motors. Pneumatic shifting of the drive transmission provided two speeds for driving, with the shifting accomplished using VEXpro CIM Ball Shifters and double acting pancake cylinders. The dual-speed feature was added to allow the robot to be moved at a slow but controlled speed when backing away from a stack of totes on the scoring platform.

A fifth drive wheel was mounted in the center of the robot orthogonal to the four corner wheels. This wheel was also an omnidirectional wheel and was raised and lowered by two pneumatic pistons. If the robot driver needed to move sideways, this center wheel would drop to contact the floor and then be driven while the four corner wheels were unpowered. The rollers on the corner omnidirectional wheels would roll in response to the sideways force generated by the center wheel, thereby translating the robot across the field. Packing so many components into such tight quarters was only possible by carefully planning the space allotment using CAD.

CAD — AN ESSENTIAL DESIGN TOOL

FRC Team 1511 embraced the use of computer-aided design, computer-aided analysis, and advanced manufacturing technologies as an integral part of a component-based design methodology. The design files produced not only the working drawings to manufacture and assemble components, but also produced a design archive of the robot. This archive can be preserved and used as a reference for future designs, a task that is much harder to achieve with a physical robot due to storage limitations and the constant temptation to scavenge components during future build cycles.

Mentors guided the process of using the embedded CAD analysis features by hand, calculating most of the parameters determined by the software. This was done to provide a reference for the computer-calculated results and to instruct students on the fundamental methods to calculate mechanical properties. Careful cataloging of parts and processes provided structure for team members to work autonomously and served as a tool to track progress. With access to all files available to all team members, individuals could integrate their work with that of others, and gain an overall understanding of how the discreet parts functioned as complete systems. This methodology kept all informed and demonstrated the value of designing with CAD tools.

◐ *The base, gripper, stabilizer, elevator, and container-grabber were fabricated using waterjet cutters, CNC mills and lathes, and traditional shop equipment, such as metal brakes, mills, lathes, band saws, drill presses, grinders, and hand tools.*

Team 1987 – The Broncobots' Conveyorbot

BRINGING THE TOTES TO THE ROBOT

Biomimicry is the application of biological solutions found in nature to solve mechanical problems. To address the 2015 *FIRST* Robotics Competition (FRC) challenge FRC Team 1987 applied a concept one might call "indumimicry" – the application of industrial solutions to solve a mechanical problem. With the announcement of the competition challenge the Broncobots, from Lees Summit, MO, turned to industrial solutions as inspirations for its design.

By reviewing commercial applications for moving storage containers, team members concluded that conveyors were by far the preferred system used to transport packaged goods. This revelation confirmed their desire to apply an industrial solution to the game challenge.

To maximize scoring the team decided to design a robotic system that brought totes from the loading station directly to the scoring platform. At that location a second system would be designed to build stacks of totes. This approach led to the creation of a composite robot consisting of two integrated systems: a conveyor system and a stacking system. The stacking system was placed on the scoring platform for the beginning of the match where it remained for the duration of the match. The conveyor provided a passive and powered path from the loading station to the stacking system on the scoring platform. The team's goal was to stack all 30 totes stored in the loading station and rely on its alliance partners to top off the stacks with recycling containers for additional points.

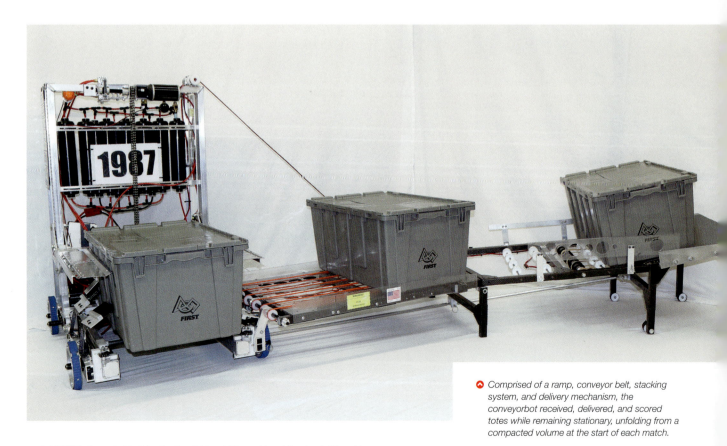

Comprised of a ramp, conveyor belt, stacking system, and delivery mechanism, the conveyorbot received, delivered, and scored totes while remaining stationary, unfolding from a compacted volume at the start of each match.

54 | *FIRST* Robots: Behind the Design | Vince Wilczynski and Stephanie Slezycki

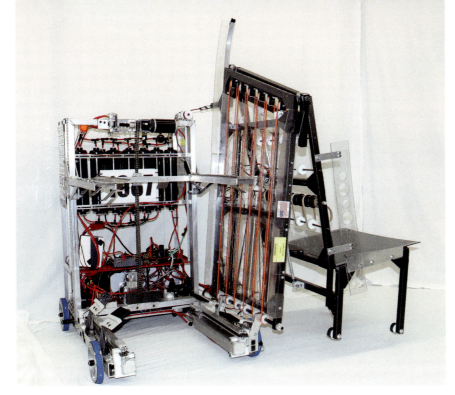

○ The unpacking sequence began by unfolding the ramp that received totes from the loading station, followed by extending both stages of the conveyor belt.

○ Planning before manufacturing ensured that the folding linkages, transport mechanism, and delivery system functioned properly.

A UNIQUE SOLUTION

Robot origami was needed to package the seven-foot-long conveyor and the stacking system into a legal configuration for starting a match. The folded configuration allowed an alliance partner to position itself in front of the loading station to pick up the autonomous totes at the beginning of the match. After the alliance partner moved away the conveyor unfolded to reach the tote loading station while the stacking system remained on the scoring platform. Once the conveyor was in place, an automated stacking routine commenced.

A series of steps transported totes down the conveyor and into the stacking system to be scored. A single tote would first fall onto the first stage of the conveyor: an elevated platform that was tilted toward the conveyor. This platform advanced the tote to the second stage of the conveyor: a powered and passive roller section. The reach of the powered roller was extended to the first set of rollers using rubber bands to grab the bottom of each tote as it exited the elevated platform. Walls along the conveyor centered the totes and prevented them from falling. The third section of the conveyor used polycord belts to advance the totes. The longest run of belt first grabbed the corner of the tote closest to the field wall. This action rotated the totes 35 degrees as the conveyor itself turned.

The totes exited the conveyor system by falling onto two carbon fiber rods that served as a retractable shelf for the stacking system. Brushes centered the totes in the stacker and absorbed some of the energy during the transitioning fall. When the tote was in place on the shelf, the elevator would be lowered, engage the tote, and lift it clear to allow a new tote to fall on the shelf.

Once as many as six totes were stacked, the elevator lifted the stack, and the carbon fiber rods that acted as a shelf retracted under the conveyor. A pneumatic cylinder and a linkage to amplify the cylinder's displacement were the mechanism that retracted the rods. The elevator would then lower the totes on the scoring platform. A pneumatic cylinder was combined with a linked set of pulleys to power a guide bracket that pushed the stack of totes 30 inches down the scoring platform. Additional stacks would in turn push prior stacks down the platform. The pushing system could generate 204 pounds of force to move an entire wall of stacked totes. Once the stack of totes was pushed out of the stacking system, the sequence was repeated to build and then place a new stack of totes.

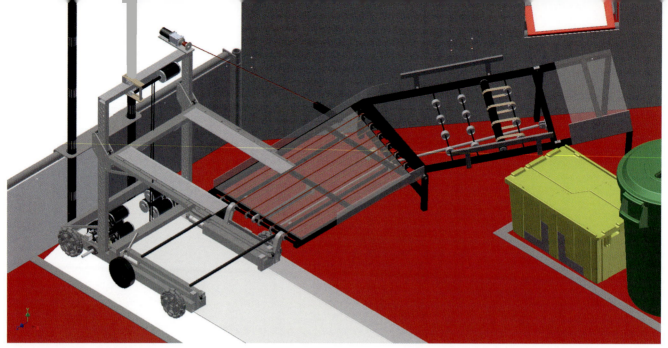

○ The robot was initially designed based on the field dimensions and the starting locations of the game pieces. Combining a robot CAD drawing with that of the field confirmed the design's parameters.

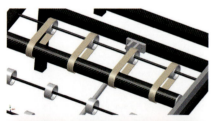

○ Many ideas were examined for transporting totes before settling on a power-driven conveyor system.

MANY ROLES FOR CAD

The conveyor design began by prototyping the concept using material already available in the team's workshop. This evolution identified the minimum angle needed for the totes to slide down a passive ramp. Equipped with this parameter the team progressed to computer-aided design (CAD) using Autodesk Inventor as its design software. CAD software was used to refine the design and establish plans to construct the robot's components.

The CAD model was used to define the angle of intersection between conveyor sections as well as the length of each polycord belt. The model also detailed the interface between the conveyor and the stacking system including the clearances needed at the hinges and the distance to retract the tote shelf under the conveyor. Carbon fiber structural components, including the powered roller and conveyor frame, were selected to conserve weight, with this detail noted on the CAD plans. The complete design was exported to a CAD drawing of the field to confirm dimensions.

The design parameters were verified based on the CAD model and a prototype of the conveyor system was constructed using aluminum to confirm the functionality of the designed system. This review of the prototype's performance included evaluating the system's ability to expand from its folded starting configuration and the ease of tote travel on the conveyor. Making improvements at this stage was important before manufacturing shifted to using the less forgiving carbon fiber structural members. With the design confirmed each CAD part was printed as a manufacturing guide to fabricate the components. Changes made by the fabricators were relayed to the CAD designers who then updated the model.

The use of CAD during this stage of the design cycle was essential to determine the interaction between all components that allowed the system to be folded

 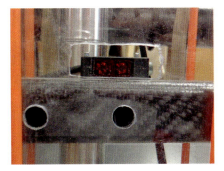 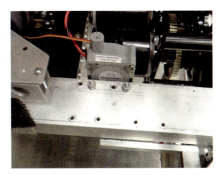

● Light sensors detected the presence of totes at various locations on the conveyor and within the stacking system. A string potentiometer measured the height of the tote lifting elevator.

into its starting configuration. Each element of the stacking and conveyor systems was evaluated to create the mechanisms and linkages to join each stage in their stored and extended configurations. The CAD model was also used to predict the robot's weight and to estimate the location of the center of gravity for the stacking system and folded conveyor. This review confirmed that the combined systems were stable and would not topple during storage or deployment.

SENSORS AND SIGNALS

Computer code written in C++ was created to autonomously control the entire process of transporting, stacking, and scoring totes using sensor data to sequence actions. As a tote exited the loading station it advanced down the conveyor, propelled by a rotating powered roller and polycords. A proximity sensor embedded between the polycords detected when the tote fell from the conveyor onto the retractable shelf. Once in position on the shelf, a second proximity sensor detected the tote's presence and triggered the elevator to lower from its default storage height. The elevator's travel was measured using a string potentiometer that was connected to the elevator carriage.

After the elevator lowered to a prescribed location it engaged the tote and then rose to its default storage height, this time carrying the new tote. It remained at this height until the next tote activated the second proximity sensor. Once six totes were loaded in the elevator the temporary shelf was automatically retracted, the load lowered, and the pushing cylinder energized to move the stack down the scoring platform. At this time the elevator returned to its default condition and the shelf extended to be in position for repeating the tote loading cycle. The entire process ran autonomously with the robot operator serving as a backup to the automated system.

Feedback internal to the control system was used to develop and debug the computer code. This included data from the controller area network (CAN) bus on the power distribution board and the smart dashboard that relayed sensor, power, and control states. In addition, the control system's data logging function was used to provide feedback with code written to print values to the log file that indicated which parts of the code were functioning properly.

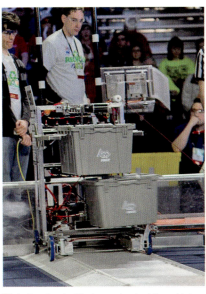

● As the elevator lifted a previously delivered tote, room was available for a newly arrived tote to be deposited in the base of the stacking system.

◗ The series of operations to transport, stack, and score totes was automated to free the robot operators from executing numerous repetitive tasks.

CREATIVE DESIGN

The team's approach to bring the totes to the stacking system was a creative and effective strategy to build and score stacks of totes. Using CAD software the team designed a collapsible conveyor system that allowed alliance play and then unfolded to conform to the field geometry. Though the robot had a drive system, this feature of the robot was not used as the stacking system did not move from its starting position on the scoring platform. Prototyping and CAD developments occurred simultaneously with the results from each step informing the other process to optimize the design.

The fact that the entire operation ran autonomously demonstrated the team's skill at monitoring and controlling each step of the process. The conveyor and stacking system were reliable and consistently produced walls of totes. Similar to the industrial systems the robot was modeled after, the only human interaction the system needed was careful design, manufacturing, and programming. At the competitions the Broncobots' conveyorbot simply ran all on its own, match after match.

◗ Industrial examples of transporting materials within factories provided models for designing a robotic system to accomplish a similar task.

THE *FIRST*® ROBOTICS COMPETITION CHAIRMAN'S AWARD

A Model Team Grows *FIRST*®

Each year, *FIRST*® charges its teams to change the culture of their local communities and motivates the teams to pursue this goal. As stated by *FIRST*, "The *FIRST*® Robotics Competition (FRC®) Chairman's Award is the most prestigious award at *FIRST*; it honors the team that best represents a model for other teams to emulate and best embodies the purpose and goals of *FIRST*. It was created to keep the central focus of FRC on the ultimate goal of transforming the culture in ways that will inspire greater levels of respect and honor for science and technology, as well as encouraging more of today's youth to become scientists, engineers, and technologists." The winner of the 2015 FRC Chairman's Award was FRC Team 597, The Wolverines, from Los Angeles, CA. As Chairman's Award winners, they join the past 22 winning teams in the *FIRST* Robotics Competition Hall of Fame.

FRC Team 597 was formed in 2001 by two teachers at the Foshay Learning Center who wanted to provide their students with hands-on learning opportunities to better prepare them for STEM careers. The students at Foshay, a Los Angeles Unified School District high school, come from predominately Latino and African American ethnic backgrounds. Approximately 82% of the students at the school are from families living at or below the nationally established poverty line, with all of the students on FRC Team 597 at or below this level.

Despite the team's seemingly limited resources, the members were determined to change their community's perception of science, technology, and engineering careers. With perseverance, teamwork, and commitment to its mission, the team has grown from just six students in 2001, to 40 students participating in the 2015 FRC season, of which more than half were female. Throughout the years, FRC Team 597 has grown as a family and takes every opportunity to SHARE its message. For this team, SHARE is an acronym for "Show How Awesome a Robotics Education is." The team members have not been held back by statistics and expectations, and are instead "SHARE-ing" their enthusiasm for science, technology, and engineering both within and beyond their community.

FRC Team 597 spreads the message of *FIRST* and raises awareness of science and technology at community events and schools, and has also had unique opportunities to give demonstrations at a comics expo, a Los Angeles Lakers game, the Los Angeles Trade-Technical College STEAM Expo, and University of Southern California football games, to name a few. Stories of their journey have been featured in the *Los Angeles Times*, *U.S. News and World Report*, National Geographic, and will.i.am's "i.am *FIRST*: Science is Rock and Roll."

FRC Team 597 is well known in its community for its extensive involvement in helping others. Each year team members are required to volunteer 200 hours of their time to community service. From running food and blood drives, as well as book and toy drives, to supporting breast cancer awareness, hosting science-based workshops for students and families, tutoring, and providing computers and instruction to needy families, there is no shortage of opportunities to volunteer.

Over the past seven years, the Wolverines have established and mentored 44 *FIRST*® LEGO® League (FLL®) teams, of which some of the current FRC team members are veterans. The students volunteer their time at FLL competitions as judges, referees, and field staff. FRC Team 597 introduced hundreds more middle school students to FLL at events such as "A Day of Robots" at the California Science Center and "I <3 Robotics Day." FRC Team 597 also supports and mentors local FRC teams, and in the past six years has helped establish five teams. It has even mentored seven Mexican teams and one Canadian team through Skype. The Wolverines share their machining facilities with other teams and host programming and robot build workshops. To help grow *FIRST*, the team created and distributed a presentation throughout the school district titled "10 Easy Steps Guide to Starting FRC, FTC, and FLL Teams in an Urban Environment."

FRC Team 597 embodies a true *FIRST* Hall of Fame team, and its success is evident. All of the students on the team are graduating high school and attending college, with 90% pursuing degrees in STEM-related fields. Their efforts have even led to the creation of a new K-12 Manufacturing and Robotics Engineering Academy for the 2015-2016 school year. Their story is inspiring — this team truly builds more than robots.

The FIRST Chairman's Award is the highest honor presented to FIRST teams. The award celebrates a team's success inspiring students to pursue education and careers in science, technology, and engineering.

CHAPTER 3
💡 INNOVATIVE DESIGN USING TRADITIONAL MACHINING METHODS

Creative thinking is the most important aspect of creative design. Knowledge and expertise are often paired with creativity to serve as the fuel for design processes that yield innovations. The transformation of an idea into an innovative solution requires that the design become a physical object that can be examined, applied, and improved. For robotics, many different manufacturing methods are generally needed to convert designs into physical components and integrated systems. This chapter showcases robots that have been produced using traditional manufacturing methods for drilling, cutting, milling, turning, and finishing. Five case studies of creative solutions illustrate that robotics innovations are independent of the manufacturing method used to create the physical manifestations of the original designs. These examples explore the thought process behind creative electromechanical design and demonstrate how manufacturing methods merely facilitate creativity to produce innovations. Modern manufacturing methods are often variants of traditional machining, where embedded control systems automate machining processes. The team profiles in this chapter showcase some of the many ways to manufacture robots.

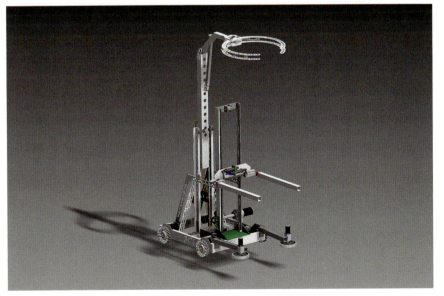

Innovative Design Using Traditional Machining Methods | 63

Team 525 – Designing Simplicity

◂ By modeling every screw, nut, rivet, structural member, actuator, motor, and power distribution component, an exact template was established and helped guide component fabrication using traditional machining methods.

CAD DETAILS SUPPORT A SIMPLE MANUFACTURING PLAN

Sometimes it takes more time to do less. This adage can apply to robot design, requiring more time to be creative and build a robot with fewer features, but greater utility and stronger performance. Cedar Falls, Iowa-based *FIRST*® Robotics Competition (FRC®) Team 525 — the Swartdogs — began the season recognizing that a simple robot was much harder to design than a complex robot. The team noted that it is easy to address design issues by adding mechanisms, but that each addition adds complexity and increases the chance for errors.

The 2015 FRC Team 525 goal was to design a clever yet effective robot that could be easily constructed using a limited collection of shop machines. To do so, the team had to be creative. A careful and consistent computer-aided design (CAD) approach and a solid understanding of traditional machining methods helped the team achieve this goal.

CAD PROMOTES CREATIVE THINKING

The team applied CAD methodology at two distinct levels. The first level involved using block geometry to establish a design envelope for the overall robot and its specific functions. The second level of its CAD methodology involved developing detailed plans to ensure that components functioned properly. This level of careful CAD documentation also allowed the team to use the CAD plans as assembly instructions during the build process. The team relied on PTC Creo for all of its CAD work.

The design process began by defining the maximum volume available for the robot to occupy. The team decided to remain within the transportation configuration defined in the 2015 FRC robot rules for the entire match. Extending beyond that configuration required additional mechanisms and violated the team's pledge of simplicity. This established a maximum-sized envelope of 26 inches wide, 42 inches long, and 78 inches tall. The first step of the design process was to carve out a volume that maximized the stability of the drive system, followed by removing an internal volume that was needed to ingest containers.

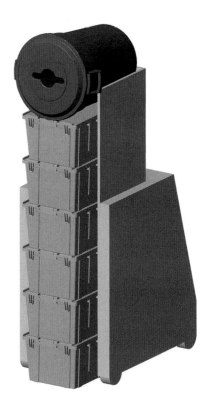

The design process began with a solid block that had the maximum volume allowed by the robot rules. The available volume for the structures and systems that would ultimately define the robot was established by removing area to accommodate each design decision. Adding the game elements to the completed block model confirmed the preserved volume for the robot's structure and confirmed that systems could meet the design goals to hold totes and a container.

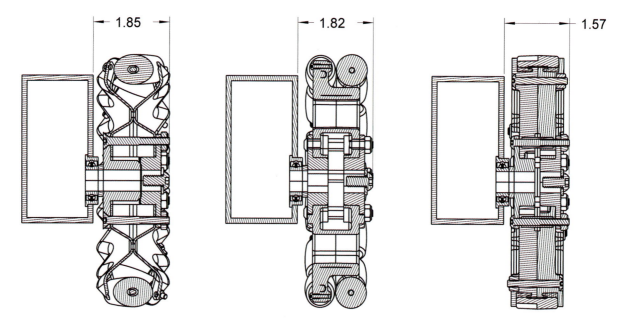

▲ The option to use mecanum, omnidirectional, or traction wheels was preserved early in the design process. A design based on using the widest option, mecanum wheels, provided a space margin for any of the other drive choices.

▲ The drive transmissions were also arranged to preserve options. For example, the transmission was designed to power two motors on each side of the robot independently, such as for a mecanum drive system, or jointly, as would be done using omnidirectional or traction wheels.

Practicalities associated with the team's primary transportation by bus to the competitions dictated an upper limit on the height of the robot's base because it had to fit in the luggage space under the bus. This limit required that additional features be modular so they could be easily removed and quickly reinstalled at each competition. The preliminary plan used this restriction and maximized the height allowed by the robot rules to enable the robot to securely transport a stack of six containers topped by a recycling container. Further analysis of this initial CAD model was used to determine operating characteristics of the design, namely the range of motion needed by the lifting carriage to pick up objects and the carriage's travel height to load totes from the loading station.

A detailed CAD model was developed within the size constraints of the initial block-geometry CAD model. As much detail as possible was included in this phase of the modeling process to ensure the integrity of the designed system. The detailed model was used to evaluate and accommodate design issues that could not be determined at that point of time in the design process. In a variety of cases during the early phase of the design process, the team decided not to decide and moved ahead by incorporating CAD configurations that preserved options.

For example, the team decided to move forward with a design that could accommodate two completely different drive systems. One design was based on installing four mecanum wheels to allow the robot to move in any direction or orientation. An alternative drive system design used two omnidirectional wheels, which have inherent low side friction, and two traction wheels to create a platform that was skid-steered. The only way to determine which system was best was to test the actual system on the robot — a condition that could only be met by accommodating for both drive options in the CAD model. As such, an accommodation for the width of the mecanum wheels was taken into account to ensure that the robot size rules were not violated.

Similarly, the transmission was designed to accommodate both drive systems. The mecanum drive option required individual motors driving each wheel, while the skid-steered option required that both wheels on the same side of the robot be jointly powered. Both of these options were preserved as a possibility by designing a transmission powered by two motors on each side that could drive both wheels simultaneously or separately. This was accomplished by either including or removing an idler gear between the motors' drive shaft gears.

The CAD drawings were created with the maximum number of features included in the plans, such as including the mecanum wheels with the idler gear even though the idler gear would not be installed in this drive configuration. By preserving the option to remove components, such as the idler gear, the flexibility of the plan increased. Regarding the drive system, the team assembled a robot base, tested the mecanum drive configuration, and determined this drive system to be effective for the competition.

The final CAD model detailed all aspects of the design including fasteners and chain. This level of detail established the components for each of the robot's subsystems including the chassis, the elevator and its frame, and the carriage that carried the containers. Each subsystem was designed and integrated into the robot first as a digital element, thereby ensuring that the systems fit and functioned as a composite unit. Each subsystem was designed with simplicity in mind so that the mechanisms could be machined in the team's shop using traditional machining methods.

EFFECTIVE MANUFACTURING

A few factors were key in the robot's fabrication. Based on the detailed CAD model, the team had a clear plan to manufacture components and construct the robot. The level of detail in the CAD model helped team members visualize the robot before it was built and while they were building the robot. Because all members of the team had access to the CAD model and the robot plans, work on each subsystem could be pursued simultaneously. Since the interfaces between subsystems were well defined in the CAD model, each subsystem team knew the operating envelope for its particular component.

The team's machine shop consisted of the following equipment: a drill press, a small manual mill, a small lathe, a chop saw, and hand tools. The robot was designed so that the students could manufacture it using this equipment. The shop equipment provided the team members with flexibility since they were not restricted by components, such as shafts, in the limited dimensions available from vendors. Rather, FRC Team 525 had the ability to fabricate custom components needed for its design. The only structural component not manufactured by the students was the 3.5-inch by 1.75-inch rectangular tube with an indented midsection used on each side of the robot. Given the unusual geometry and the required accuracy for locating the wheels, sprockets, and gear shafts, this component was manufactured by one of the team's supporters in a commercial shop.

◐ Benchtop tools including a small mill, lathe, and drill press, as well as power tools, were used to fabricate parts of the robot.

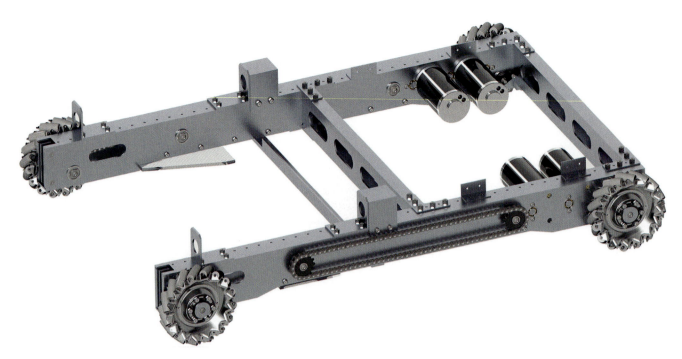

↑ The CAD version of the robot included variations of systems that could be accommodated within the available space. For example, the power transmission was designed to support three versions of a drive system, with the physical robot then configured using only those components needed for the chosen system.

Material selection was important in the design process. An early version of the robot frame was constructed from one-inch square aluminum tube stock having a 0.1-inch wall thickness. This light duty material was easy to cut and handle. After carefully fitting the pieces to be square, the VEXpro Tube Stock and Gussets were riveted together. Given the integrity of this component, a revised version was later constructed using one-inch square tubing with a 0.04-inch wall thickness. Incorporating this change into all of the elevator supports, except for the front two vertical pieces, retained functionality and reduced the weight by eight pounds.

The value of the detailed CAD model was also realized in the design and fabrication of the drive system for the elevator. Each corner of the cradle holding the containers was attached to a chain that raised and lowered the carriage. With one motor turning a common shaft, the lifting sprockets on each side of the robot rotated in unison and kept the carriage level. Cleverly, the two motors were connected to each other using sprockets, chain, and a pair of gears. In the case of uneven loads on one side of the cradle, this linkage lifted the carriage in a level manner since the gearing ensured both motors rotated at the same speed. The motors and upper chain sprockets were mounted using VEXpro VersaFrame Parallel Mounts — an easy-to-use and reliable method to mount hardware and align shafts. Care was taken during the assembly process to properly align the lifting chains to avoid binding.

Simple construction and assembly methods were also apparent in the frame fabrication and transmission installation processes. Knowing the importance of a level frame to ensure that the mecanum drive system would function properly, the team repurposed a door to provide a flat and level surface in the shop. This surface became a worktable to install the drive system. Slots were milled in the frame's cross members to facilitate assembly and reduce weight. The CAD plans served as an inventory and assembly reference to collect the transmission components and install every gear, shaft, sprocket, bearing, and spacer into the frame.

The cradle was also fabricated using the CAD plans for parts inventory and as assembly instructions. The

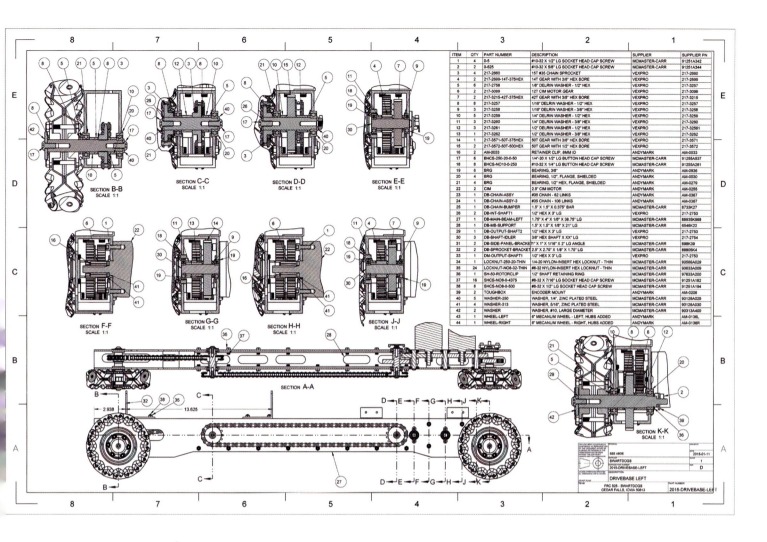

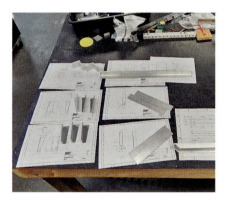

The CAD plans served as templates for the assembly process. The drawings became placemats for all of the components needed to build each structure, mechanism, and system.

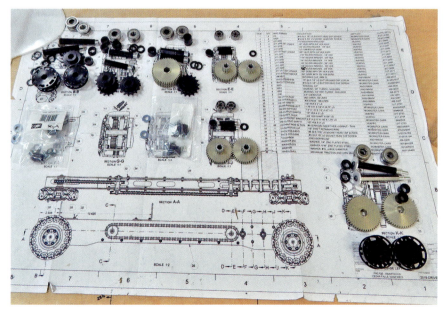

Innovative Design Using Traditional Machining Methods | 69

◐ ◑ *The robot was designed to allow fabrication to be completed using screws and rivets, with the location of each fastener accounted for in the CAD model. Provisions were also provided to access each fastener, with the required accessibility machined into the parts during the fabrication process.*

part drawings were distributed to the student fabricators in the shop, with the students then delivering the parts and drawings to the students who assembled the cradle. Three iterations were made on the cradle design based on performance tests to optimize polycarbonate fingers that flexed inward when coming down over the top of a container and had the rigidity to lift the container when moving upward. Each design iteration was documented as a CAD model before being fabricated. A simple manufacturing and assembly process decreased the time required to test ideas and evaluate changes.

THE VALUE OF SIMPLICITY

The team's performance in the 2015 FRC season — winning two regional competitions and finishing in the top 10% of teams at the *FIRST* Championship — validated the effectiveness of its design methodology. The design process was one of openness and inclusion with all team members aware of the process and engaged in finding solutions to challenges. One example of the value of this open communication was a suggestion from a member of the build team who recommended running the center section of the drive chain on the outside of the frame. This idea was not obvious to the design team, but the change greatly improved the ease of assembly for this section of the robot.

One tradition was especially noteworthy as it reflected the winning attitude of FRC Team 525. Just before the robot was completed on the last day of the build season each member of the team engraved his or her name on the robot. It was a small, simple gesture that spoke volumes of each person's ownership in the design and fabrication process.

FRC Team 525 was guided by its mentor's advice: "The role of a design engineer is not to create great designs but to get great designs into production." A great design was put into production as the FRC Team 525 robot — a product that exemplified the value of a simple design that was carefully planned, fabricated, and assembled.

CAD was used to plan every aspect of the robot, including the placement of each sponsor's logo. This level of exactness in planning eased the fabrication and assembly processes where the CAD plans provided instructions for each step.

Team 2481 – Strategic Use of Machining Resources

DESIGNING AN ACHIEVABLE AND COMPETITIVE ROBOT

Robot design can be influenced by many factors: some teams may focus on game strategy while others design around resource limitations. There are also teams that can not only accommodate both, but excel at integrating strategic robot design with limited tooling and material. FIRST® Robotics Competition (FRC®) Team 2481 — the Roboteers — are one such team. Its creative efforts resulted in an award-winning robot that distinguished the team among the other teams.

ANTICIPATING THE GAME

Upon learning the game specifics, the Roboteers began the design process by ensuring all team members thoroughly understood the competition rules and robot regulations. Once all team members were familiar with the game, they began assessing potential strategies and robot designs. A full-sized playing field was taped onto the school gymnasium floor and the students assumed the roles of robots they might expect to see in competition. This activity enabled the team to directly experience time and space constraints, the physical properties of the totes and recycling containers, and the interaction between robots. This hands-on activity helped the team anticipate the types of robots that could be built and identified the most challenging aspects of the game. FRC Team 2481 identified two potential robot designs differentiated by how they would collect totes — from the loading station or from the landfill. The students also speculated about which type of robot other teams would build and how their own robot would interact during game play.

To determine which of these robot types the team should build, the students surmised that most teams would attempt to build a robot to collect totes from the loading station because this would require a less complex intake design to accommodate the predictable tote orientation. Thus, the Roboteers decided to pursue a landfill robot, which would be more difficult to design and execute, but would also better complement other teams as an alliance partner.

The game simulation also helped the team members explore scoring scenarios, and they determined that the recycling containers had the most potential to influence the outcome of each match. The alliance that gained control of the recycling containers in the center of the playing field would have the scoring advantage. The team also reviewed human player tote loading rates, tote quantities, and pool noodle placement in the recycling containers.

As each competition season progresses, the manner in which teams and their robots play the game also tends to evolve, as different strategies emerge and teams better understand alliance interactions and refine scoring priorities. The Roboteers spent some time reflecting on this evolution, from qualification

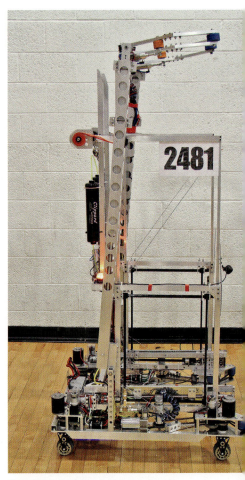

○ FRC Team 2481's robot was designed to collect totes and recycling containers from the landfill, thus freeing up the loading station for alliance partners.

◐ The initial arm design was prototyped out of stock aluminum pieces and fit to a preliminary chassis to verify mounting points and alignment with the totes.

rounds to playoff rounds at regional competitions, and even how the game would be played at the Championship. The team anticipated that a robot's design may need to be modified to stay competitive. The Roboteers adopted a competition philosophy that would drive future design decisions: that a good robot with great strategy will always outperform a great robot with mediocre strategy.

PRACTICAL FUNCTIONALITY

FRC Team 2481 designed a robot that could perform a three-tote autonomous stack, deliver a stack of six totes topped with a recycling container to the scoring platform in teleoperated mode, and also cap its alliance partners' stacks up to five totes high. After preliminary whiteboard sketches, the virtual design process began and the robot components were drawn using Autodesk AutoCAD, a computer-aided design (CAD) software application. This enabled the team to visualize what parts were needed, how to actuate moving components, and how to best route pneumatic tubing and electrical wires.

The Roboteers identified that mobility would play an important role in this game and selected a drivetrain to satisfy this need. A swerve drive with small diameter, high traction wheels was incorporated to maneuver the robot in confined spaces and to navigate the elevated scoring platform. Totes were collected from the landfill when they contacted the robot's compressible intake

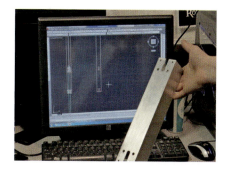

◐ The arm was modeled using AutoCAD software prior to the start of fabrication, which helped the team incorporate important design features prior to reducing the weight of the components.

Innovative Design Using Traditional Machining Methods | 73

An end mill was used to cut slots and holes in the arm for bearing roller clearance and mounting brackets.

The inner telescoping arms were actuated by pneumatic cylinders and were guided along roller bearings. A pivoting wrist at the end of the arms correctly positioned the attached claw for grasping recycling containers.

wheels, which aided in aligning and securing the totes into the robot's internal indexer to be stacked.

The defining feature on the robot was the telescoping arm assembly used to acquire recycling containers from the step in a controlled manner. The assembly was designed to pick up the containers from any orientation and position them above the totes in the indexer to cap each stack. The arms also served to stabilize tote stacks during transport to the scoring platform, and to cap stacks built by alliance partners. The team calculated that the arms needed to extend an additional 18 inches over the landfill totes to reach the recycling containers.

Each extending arm was actuated by an 18-inch stroke pneumatic cylinder and the 90-degree pivoting function of the assembly was driven by a motor. The arm assembly also required a motor-driven wrist with a 180-degree pivot range to correctly position the claw to the desired orientation. The claw used to grasp the recycling containers was actuated by two 12-inch pneumatic cylinders, and the ends of the claw featured BaneBots high grip wheels to maintain control of the containers. The choice of a swerve drive eliminated the need to add sideways motion to the arm assembly.

DESIGNING FOR CAPABILITY

Throughout the robot design decision-making process, the team maintained an awareness of its limited resources for fabrication. The robot was manufactured in the former school automotive shop, outfitted with a manual milling machine with X-Y-Z readouts, horizontal and vertical band saws, bench grinders, a one-ton press, table and miter saws, a drill press, and various handheld tools.

The available tooling required significant labor and attention to detail. The telescoping feature of the arm assembly consumed the largest amount of design and manufacturing time. Over 11 hours were spent machining the arm assembly's four pieces of aluminum to meet strength and weight requirements. Understanding its resources, the team was prepared to take on the laborious task knowing it would eventually manufacture a quality robot component. Robot fabrication materials were also selected to align with the shop's limited capability. The team determined that aluminum exhibited the desirable mechanical properties for its robot, including weight, rigidity, machinability, and availability in standard sizes.

The machining of the robot's aluminum components began with a band saw to cut the pieces to length. These rough-cut pieces were then end milled square to the longitudinal axis of the material, providing a reference point for the remainder of the machining operations. The longitudinal axis was then zeroed to the machine, and the radius of the cutting tool was added as an offset. Once verification had been made that the stock material was clamped in the mill vice correctly, the piece was end milled to the specific dimensions needed. The team's careful attention to detail and adherence to tight machining tolerances also helped fabricate spare parts that were perfectly interchangeable with the original pieces.

Paper printouts were used to locate functional holes and slots on the arms prior to removing additional material for weight reduction. Various-sized end mills were used for most of the fabrication, while a bore was used for holes that required stricter tolerance. All machining on the robot was planned around available tools, including the inability to mill square corners with the rotating cutting tools.

Robot design continuously evolved to maintain the desired functionality, such as resizing slots to provide sufficient corner clearance, while working within machining limitations. Occasionally, the functionality requirement of a part was deemed important enough that a redesign was not the optimal choice. The outer sheath of the telescoping arms, constructed from aluminum tubing that measured two inches wide by two inches tall and 0.125 inches thick, would need a series of large 1.5-inch diameter holes to reduce weight, so the decision was made to purchase an end mill large enough to satisfy the requirement.

Another example of designing within capability constraints occurred with custom milled aluminum angle brackets required to mount the drivetrain components. An angle stock leg length of two inches or greater would interfere with clearance to the milling

A large end mill was purchased specifically for machining the 1.5-inch diameter weight reduction holes in the outer sheath of the telescoping arms. The machining was performed by students on a manual milling machine with X-Y-Z readouts. Many hours were dedicated to these structural components to ensure they met strength and weight requirements.

⬤ The telescoping arm assembly extended an additional 18 inches, reaching over the landfill totes to grab recycling containers from the step. This provided additional opportunities for FRC Team 2481's alliance to cap tote stacks for additional points.

chuck and the cutting tool, or with the top of the mill table. The Roboteers identified three possible solutions: to redesign the part, fabricate a custom fixture to hold the piece for machining, or purchase longer cutting tools. Due to time and financial constraints, the team chose to redesign the brackets.

FRC Team 2481's design approach and practical execution of robot fabrication earned the team the Industrial Design Award (sponsored by General Motors) and the Quality Award (sponsored by Motorola) at two regional competitions. This team was able to build a competitive robot while making provisions for known resource limitations by eliminating complicated parts and minimizing complex manufacturing operations. The Roboteers' innovation can be found in the team's ability to devise creative solutions through inventive thinking and present these solutions in the form of a successful FRC robot.

⬤ The robot lifted a recycling container from the playing field floor during a match in preparation to cap the totes collected from the landfill.

Team 2485 – Spotlight on Carbon Fiber

APPLYING COMPOSITES TO ENHANCE DESIGN AND PERFORMANCE

Carbon fiber-reinforced plastic, also referred to as simply carbon fiber, is an extremely strong and lightweight material. Strands of thin carbon filaments are bundled then woven into a fabric which, when combined with a resin, creates a solid structure with a high strength-to-weight ratio. One creative group of students and mentors that has created an identity with this composite material is *FIRST*® Robotics Competition (FRC®) Team 2485, the W.A.R. Lords (We Are Robot Lords), from San Diego, CA.

This team began using carbon fiber on its robot in 2011 when it welcomed a mentor with experience in composites. Initially, carbon fiber was used for aesthetic appeal purposes only, but beginning in 2014, it was applied to structural components for overall robot weight reduction. For the 2015 competition, team members used two distinct methods of creating custom carbon fiber parts to reduce component weight, increase strength and stiffness, and of course, to look cool. Carbon fiber has become part of the team's trademark look and continues to be used in new applications.

DUAL-ELEVATOR DESIGN

To be competitive in RECYCLE RUSH℠, team members knew they had to design a robot that was skilled at manipulating both totes and recycling containers. One of the biggest design challenges was related to the large volume of these two game pieces relative to the size of the robot. The team decided to pursue a dual-elevator system, one for each type of game piece, with a ratcheting hook to hold tote stacks close to the robot's frame.

The elevator system for the totes was supported by a structural frame referred to as the "strongback." This frame was connected to a lead screw that controlled the angle of the support and tote stacks relative to the ground. Stacks were essentially leaned back for stabilization during transport. A claw for manipulating the recycling containers rode up and down a central elevator shaft. Totes were drawn in to the robot through pneumatically actuated intake arms with powered belts. Once collected, totes were lifted by these arms and placed on a ratcheting hook. Once the tote was secure on the hook, the arms would lower to collect another tote. This process was repeated until a six-stack of totes had been built, with the lowest tote always supported on the hook. The robot also had a mechanism for grabbing recycling containers from the center step during the autonomous period.

FRC Team 2485 used computer-aided design (CAD) software SolidWorks to model the robot. Rough block shapes were created to test robot configurations and interaction with the game pieces. The team was able to virtually experiment with different concepts to determine the best method of manipulating both totes and

Within six weeks a SolidWorks model of the robot quickly materialized into a machine ready for competition. A dual-elevator system was efficient at manipulating both totes and recycling containers.

◐ ◑ Carbon fiber is a strong and lightweight material composed of resin and bundled strands of thin carbon fiber. The high strength-to-weight ratio makes this composite an attractive replacement for heavier structural components. FRC Team 2485 has embraced this manufacturing method, which has become part of its trademark look.

recycling containers. Prototypes of the major mechanisms, such as the belt-driven intake and the tote hook, were built. Once proof of concept work was complete, the rough computer model was refined to create the finalized robot components. GrabCAD Workbench file management and collaboration software was used to organize drawing files and monitor revision history.

THE SHOP AND MAKERPLACE

FRC Team 2485 built its robot, Valkyrie, at the school in a room affectionately referred to as "the shop." Here the team had access to a band saw and drill press, a chop saw, a computer numerical control (CNC) mill that was used to make complicated brackets and plates for the robot, and a metal inert gas (MIG) welder, which students used to create the robot frame. Team members also had access to additional tools and machines, including a laser cutter and metal lathe, at MakerPlace, a membership-based workspace and do-it-yourself (DIY) fabrication facility located down the street from the school.

TWO MANUFACTURING METHODS TO CREATE CARBON FIBER COMPONENTS

The carbon fiber on Valkyrie was manufactured using two distinct methods, referred to as wrapping and molding. The team used the wrapping method with 0.06-inch wall thickness aluminum tubing. The process began with sanding and buffing the aluminum piece in preparation for being wrapped in two layers of carbon fiber fabric. The tubing was coated with a thin layer of glue and then wrapped with the fiber, with additional layers included in the areas that would experience high stress loads. An epoxy resin was then applied to the fiber to harden the carbon, ensuring that the entire piece was fully saturated. Before the epoxy dried, the part was wrapped in Peel Ply, a fabric treated with a release agent. This was used to fill in any holes and to give the carbon fiber a flat texture. The epoxy was left to set overnight, and, when dry, the Peel Ply was removed and the piece was sanded until smooth. To achieve a gloss effect, the part was coated with a clear paint.

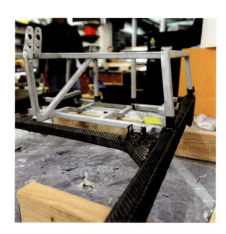

◐ The aluminum tubing of the drive base was wrapped in carbon fiber and coated with an epoxy resin. When dry, the pieces were sanded and coated with a clear paint.

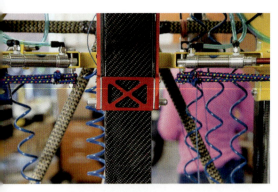

◉ The central elevator shaft supporting the recycling container claw was wrapped in carbon fiber. Behind it can be seen the Kevlar-wrapped structural supports of the strongback.

◉ The custom curved shape of the recycling container claw was created from a piece of foam cut to shape, then wrapped in carbon fiber.

Molding was used to create new parts, rather than applying a cover to a preexisting component. A mold for the custom piece was machined out of Delrin on the CNC mill to form the outer surface. The core of the part was made from foam cut to shape and then sanded. Pre-cut sheets of carbon fiber fabric were pressed into the mold, which had been coated with Frekote, a mold release agent. The quantity of layers and orientation of the fabric weave pattern were adjusted depending on the strength and stiffness properties required for the final component. The core was then placed inside the mold and wrapped with the carbon fabric. Epoxy was applied and the mold with the carbon and foam inside was allowed to dry overnight. Team members were careful to remove any air bubbles, which could reduce the strength of the final component. When the epoxy was set, the piece was removed from the mold, then sanded, painted, and installed in the robot. The team also used Kevlar for components that might encounter substantial impact during the competition matches.

Carbon fiber has many properties that make it a preferred material for certain applications, but there are also tradeoffs to consider. The material is costly, the process is labor intensive, and a certain level of skill is required to produce quality parts. The team prepared students for this by including a lesson on carbon fiber in the training workshops for new members. Under load, carbon fiber will not yield and once it reaches its ultimate strength, will experience sudden and catastrophic failure. This must be considered when designing parts, so the team used SolidWorks simulations to test the tensile and yield strengths of the robot's components.

CARBON FIBER APPLICATIONS ON VALKYRIE

The claw used to pick up recycling containers was made from a foam core wrapped in carbon fiber. The ratcheting hook that was used to hold totes under the handle had also been made with this method, as well as embedded with aluminum rods. After testing, team members found that the carbon had cracked, which reduced the structural integrity of the part and caused the hook to no longer fit under the tote handle. To eliminate this problem the piece was manufactured from solid aluminum on the CNC mill, which was produced quickly, in part because the design had already been modeled using SolidWorks software.

Although this was the W.A.R. Lords' largest robot to date, it was also the team's lightest. Weight was monitored from the beginning of the CAD process using the software's integrated weight estimate tool. One area identified for weight reduction was the strongback. Instead of using 0.09-inch wall thickness aluminum tubing, team members built it from tubing with a 0.06-inch wall thickness wrapped in carbon fiber. The two versions of the structure were compared with the SolidWorks simulation tool and showed little difference in strength, even though the carbon fiber-wrapped tubing weighed 16 pounds less than the original design. This same carbon fiber and aluminum tubing construction was used on the drive base.

FORM AND FUNCTION

While carbon fiber wasn't always necessary to achieve the intended functionality of robot components, FRC Team 2485 has been incorporating it for years and it has become a part of the team's design aesthetic. The incorporation of carbon fiber into the structure of this robot helped earn FRC Team 2485 the Excellence in Engineering Award sponsored by Delphi at the San Diego Regional competition.

What started off as a feature intended to illustrate the team's pride in its robot evolved to play a strategic role in the robot's functionality. Composites were invaluable in reducing the weight while simultaneously increasing the strength of Valkyrie's key components. As team members of the W.A.R. Lords continue to expand the integration of carbon fiber into their robots, they will reinforce their trademark image across the FRC community.

Carbon fiber is a part of FRC Team 2485's design aesthetic and was a prominent feature on its robot, Valkyrie. The composite material enabled increased component strength in parallel with weight reduction, two design features that contributed to the robot's success.

Team 3339 – Achieving Excellence with Basic Tools and Design Ingenuity

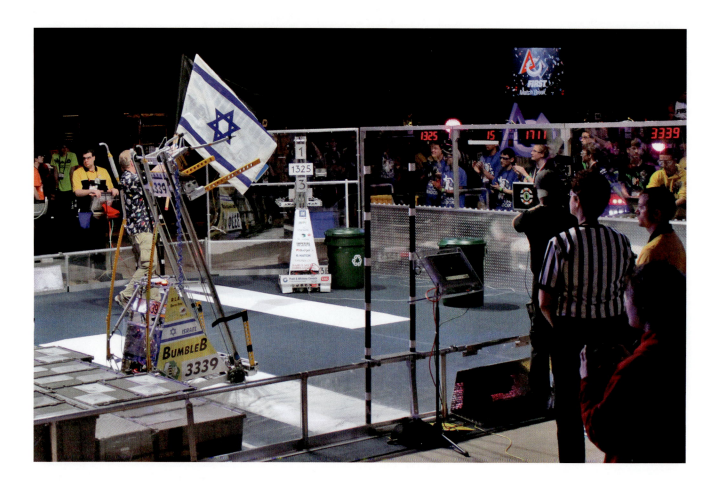

FRC Team 3339 reached a new milestone in 2015, becoming the first Israeli team to play on the Einstein field at the Championship. The students proudly stepped into the spotlight, as their robot flew the Israeli flag for the first time on the prestigious field.

ISRAEL BUZZES ONTO EINSTEIN

Students from 19 different countries participated in the 2015 *FIRST*® Robotics Competition (FRC®). While the United States and Canada contributed the majority of the 2,900 total competing teams, Israel holds the record for most teams per square mile. Since this country's first involvement in 2005, it quickly grew to 55 teams competing in RECYCLE RUSH℠. FRC Team 3339, BumbleB, had an exceptional season, reaching a new milestone for its country. While this team impressively won the 2015 Israel Regional and was ranked number one in qualification rounds, its crowning achievement came at the *FIRST* Championship. As a member of the winning alliance in the *FIRST* Championship Carson Subdivision, BumbleB became the first Israeli team to play on the Einstein field. Hailing from Kfar Yona, a small city less than five miles east of the Mediterranean

82 | *FIRST* Robots: Behind the Design | Vince Wilczynski and Stephanie Slezycki

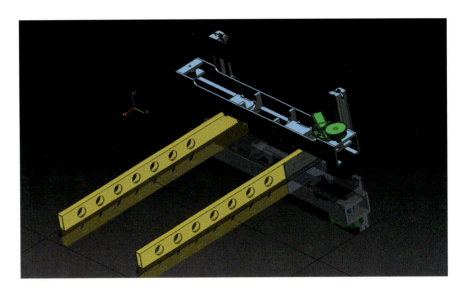

⬆ ➡ The tote arm subsystem was modeled in Siemens NX software and is depicted both with and without the decorative cover. The sliding carriages and motor adjusted the width of the arms to collect totes in different orientations. The models were used to identify system integration issues early in the design phase.

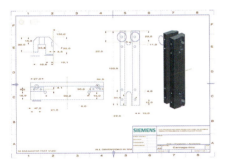

⬆ Two-dimensional drawings of the tote arm sliding carriages were printed from the Siemens NX model. The drawings were used by students as construction guides to fabricate the components in the workshop.

Sea, these students stepped into the spotlight to show the world what they were capable of achieving.

FRC Team 3339 set out to build a robot for RECYCLE RUSH℠ that could not only independently score points, but also assist alliance partner robots with building stacks. The robot, which the team members named "Clark," consisted of six subsystems: the chassis, tote arms, aligning arms, container claw, track and tilt mechanism, and container grabbers. These components enabled the robot to collect, raise, and stack totes and recycling containers from any orientation. The robot was also equipped with reliable arms to grab recycling containers off the step, and had a unique feature to evenly distribute the weight of a full tote stack across the drivetrain.

CAD HELPS DETERMINE MANUFACTURING METHODS

FRC Team 3339 used Siemens NX software as its detailed robot design tool. Once the team members had selected a strategy and finalized the robot concept, rough dimensionless sketches were created using the software. The size of the robot was then determined by the dimensions of the playing field and by identifying strength requirements of each robot component. Students developed computer models of the robot subsystems, working closely with the build teams to make updates as testing progressed. The well-developed subsystem models were then assembled to create a main robot assembly to identify any potential interferences. This system integration process identified problems early in the design phase, thus saving the team time and money by preventing errors. The final model was then used as a guide to build the different robot subsystems.

Each robot component was evaluated to determine the best fabrication methods and tooling to use. The team then used the models created in NX to print two-dimensional (2D) drawings to use as guides for construction. The majority of the robot was fabricated by the students at the team's workshop,

> Totes could be collected from both the landfill zone and loading station. A set of aligning arms mounted to the base of the robot correctly positioned totes, which were then lifted by the tote arms to build stacks.

> Team members worked to assemble the pneumatic cylinders that tilted the track and top four totes of a six-stack.

which was equipped with hand and power tools, including a jigsaw, bench grinder, drill press, and band saw. The students who participated in shop use and safety training were allowed to use most of these tools under mentor supervision. Occasionally, a robot part required fabrication using more complex manufacturing methods and was sent out to the team's sponsors for assistance. Gal-Kifuf, a metalworking company, provided both material and machining labor hours to the team, particularly with computer numerical control (CNC) cutting, punching, and bending. The team sent portable document format (PDF) drawings of the component to be manufactured, as well as a drawing exchange format (DXF) file of the flattened part to the lead engineer at the company, who reviewed the files and approved them for fabrication. Machining work was also provided by R.L.A. Development, a sponsor that aided the team with milling, lathe work, and aluminum welding. Mentors and student team members brought drawings of the parts they needed assistance with and often helped perform the machining.

SLIDING CARRIAGES FOR FLEXIBILITY

BumbleB wanted the robot to have flexibility when collecting and stacking totes, especially considering the alternating tote orientations in the landfill zone. The tote arms were attached to sliding carriages that adjusted the arm width for collecting totes from a narrow or wide side approach. The custom carriages rode up and down a vertical track, and the roller chain-driven system was raised and lowered by a CIM motor with an AndyMark WormBox gearbox. The tote arms collected totes from either the landfill zone or loading station

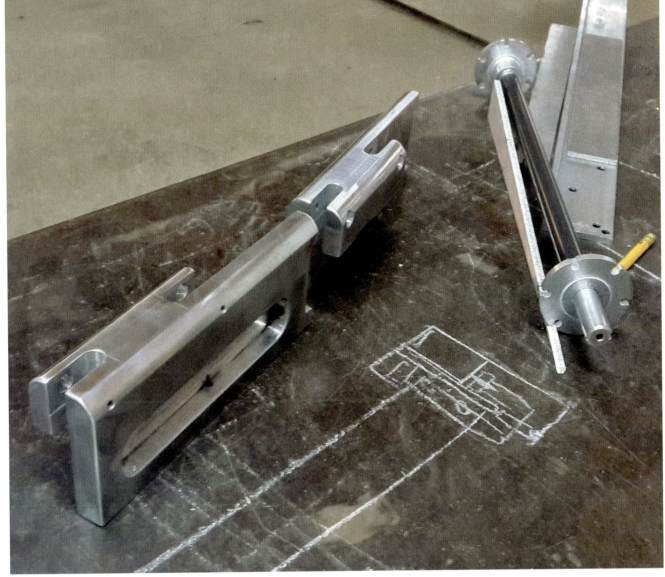

A pair of aligning arms correctly positioned the totes for the tote arms to lift. They were attached to sliding carriages, machined on the mill. A custom shaft was also manufactured to support the tilting feature of the vertical track.

via an indexing motion. As each tote was lifted and another appeared underneath, the arms slid down the track and grasped the next tote, repeating until a full stack was made.

A set of aligning arms was mounted to the base of the robot to correctly position totes for the tote arms. The aligning arms were fabricated from rectangular aluminum tubing, with bent strips of aluminum mounted perpendicularly at the ends. Each aligning arm was attached to a sliding carriage made on a mill. The aligning arms extended outwards when the robot approached a tote, then were pulled together to properly grasp and align the tote. The extension was large enough to accommodate totes in either a narrow or wide orientation. When collecting from the loading station these aluminum strips guided the first tote to land in the correct position, then aligned subsequent totes fed through the chute before being lifted by the tote arms. Upon delivering totes to the scoring platform, the carriages moved to open the aligning arms. As the robot backed away, a pneumatic cylinder pivoted the arms upward to prevent interference with the tote arms and container claw.

⊙ The container claw was used to lift recycling containers and place them on top of tote stacks. An intricate welded linkage allowed the opening and closing of the claw to be actuated by a single pneumatic cylinder. Even with six totes below, the claw maintained contact with the base of the container, providing stability during transportation.

⊙ The container claw could upright a knocked over recycling container by pivoting it on a pool noodle attached to the tote arm connecting plate.

SIMPLE SOLUTIONS FOR RECYCLING CONTAINER MANIPULATION

A mechanism called the "container claw" was used to pick up recycling containers from an upright position, hold them securely on top of a stack built by the robot, and to cap existing stacks. The robot used a feature attached to the tote arms to right containers that had been knocked over. A piece of pool noodle was attached to the plate that connected the two tote arms. When the tote arms grabbed a fallen container, the noodle caught under the recycling container lid, acting as a pivot point to rotate the container back to a standing position. The claw was mounted onto carriages that rode up and down the same track as the tote arms, with power delivered from timing belts driven by a MiniCIM motor and WormBox gear train. The opening and closing of the claw was actuated by a single pneumatic cylinder through an intricate linkage that was welded. Once the claw had positive control of a container, it was raised to the top of the track. As totes were stacked below, they supported the weight of the container as it was pushed up through the claw. With a full six-stack of totes, the claw reached around the very bottom of the container and provided stability while transporting to prevent it from falling off the stack.

BumbleB wanted to capitalize on the potential points offered by the recycling containers positioned on the step at the beginning of each match. To do so, the robot incorporated recycling container grabbers made of two long arms of square aluminum tubing with sections of round aluminum tubing near the ends. The extension and retraction of each of these rapidly deploying arms was actuated by pneumatic cylinders. At the start of the autonomous period, they forcefully extended towards two containers, giving the robot an advantage over opponent robots attempting to collect the same containers. Once the arms engaged inside the holes in the lids of the recycling containers, the robot retreated towards the loading station, bringing the containers with it. At the end of the autonomous period, the cylinders retracted, raising the arms back to a vertical position. To ensure a clean release from the recycling container lids, the end of each arm had a flexible pivot point in the round

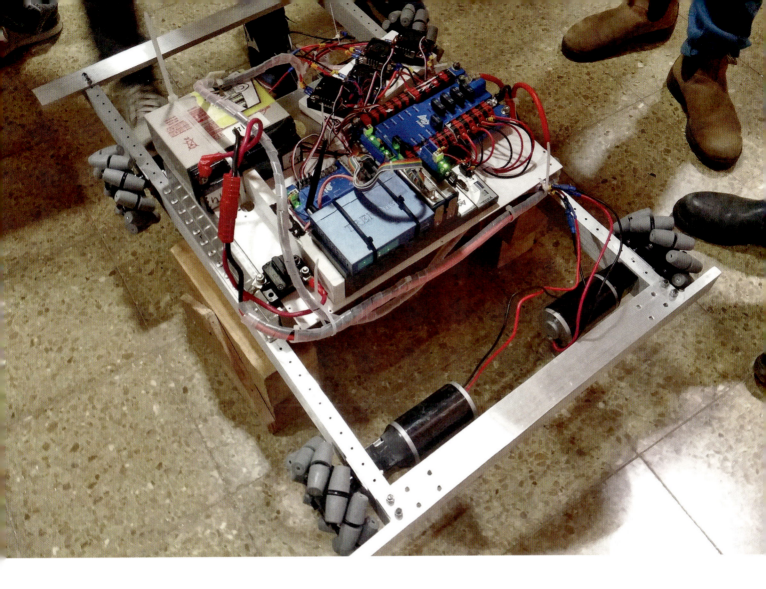

◈ A preliminary electronics board was mounted to the chassis and drivetrain for initial system tests. This allowed the team to work on programming and troubleshooting before the robot was completed.

tubing to prevent the arms from getting snagged in the lids. At the Israel Regional competition, the robot used these grabbers to collect two containers in each of the 11 matches that the mechanism was deployed, from qualification matches through finals, with a 100% success rate.

CONSIDERATION FOR WEIGHT DISTRIBUTION

The tote arms and container claw traveled up and down on a common track structure made from unique L-shaped aluminum tubing donated by sponsor David Dabi, a manufacturer of aluminum and glass products. The base of the structure was reinforced with an aluminum rod to reduce the likelihood of the track bending from repeated lifting of heavy tote stacks. Axles positioned at the top and bottom of the track supported the sprockets and timing belt pulleys, which freely rotated on the axles, allowing the tote arms and container claw mechanisms to raise and lower independently.

A highly maneuverable drivetrain that utilized four independently powered mecanum wheels gave the robot the ability to drive in any direction without

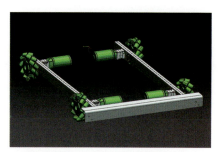

A tilting function was added to shift the weight of the top four totes in a six-stack, increasing stability during transportation. Two pneumatic cylinders controlled the tilting action, returning the six totes to a vertical orientation upon delivery to the scoring platform.

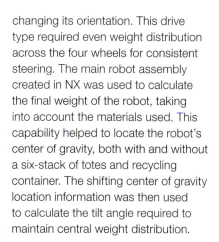

A three-dimensional model of the drive system was created to assist integration with the other robot subsystems. The frame, motors, mounts, and mecanum wheels were included in the model.

changing its orientation. This drive type required even weight distribution across the four wheels for consistent steering. The main robot assembly created in NX was used to calculate the final weight of the robot, taking into account the materials used. This capability helped to locate the robot's center of gravity, both with and without a six-stack of totes and recycling container. The shifting center of gravity location information was then used to calculate the tilt angle required to maintain central weight distribution.

For transportation of the heavy six-stacks of totes, the team designed a tilt mechanism to shift the weight of the upper four totes. Two pneumatic cylinders mounted to the top of the robot chassis controlled the angle of the track. At full extension the track was vertical; upon retracting, the cylinders caused the track to tilt towards the center of the robot. The lower two totes of the stack were guided by the aligning arms as the robot maneuvered. When the robot reached the scoring platform, the cylinders were again extended to bring the track back to vertical, and a balanced stack of four totes was placed on top of the lower two totes.

BUMBLEB'S INTERNATIONAL SUCCESS

FRC Team 3339 strongly feels that Clark was the team's best robot since its 2010 rookie year. Unlike many teams that constantly tweak, adjust, and change robot features throughout the competition season, BumbleB didn't feel the need to adjust any of the robot's systems. Team members established a solid robot design early on and were able to build exactly what they wanted using the available traditional manufacturing methods. Since they felt no need to change their strategy, the robot that performed so well at the Championship was identical to the form it took on the last day of build season. The team's design impressed judges as well, earning it an Excellence in Engineering Award sponsored by Delphi at the Israel Regional. 2015 was an unprecedented year for both the team and Israel, an accomplishment in which BumbleB takes tremendous pride.

◯ Two rapidly deploying arms, actuated by pneumatic cylinders, quickly and reliably grabbed two recycling containers from the step during the autonomous period. This arm design had a 100% success rate at the Israel Regional competition.

Innovative Design Using Traditional Machining Methods

Team 3478 – Robot Redesign: Doing a "180"

Prototypes were made using recycled parts from past robots to help refine the final design. A mockup of a working recycling container gripper and four-bar linkage was created from scrap pieces of aluminum.

CLEVER REPURPOSING OF ROBOT COMPONENTS

The design and development process for any *FIRST* Robotics Competition (FRC®) team is a critical part of the six week build season. The team makes important decisions that it will commit to until its last competition. Often, as time progresses, unforeseen challenges arise, or improved versions of past ideas are revealed. Robot redesign can be a risky endeavor, even more so later in the season. FRC Team 3478, LamBot, from San Luis Potosí, the capital city of the Mexican state of San Luis Potosí, experienced the need for redesign after its first regional competition. The team was prepared and was able to clearly identify the desired changes and implement a solution with minimal overall redesign. By keeping the original mechanisms and creatively repurposing them, LamBot was able to overcome the challenge.

SETTING GOALS TO GUIDE ROBOT DESIGN

The design process for FRC Team 3478 included a review of the important rules and an analysis of the game pieces. The team then began identifying strategies and robot mechanisms that were essential for success in the competition. Goals were set for performance in both the autonomous and teleoperated periods, and team members began developing solutions to achieve these goals. The preliminary goals included the ability to stack three

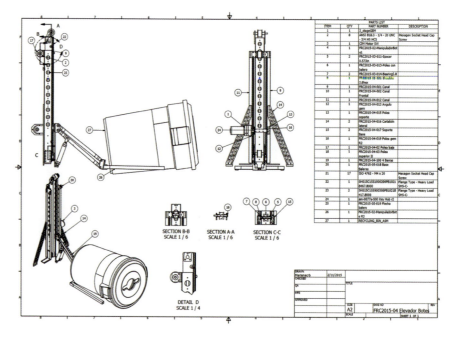

Once the final robot concept was established, each component was individually designed using Autodesk Inventor software. This modular approach allowed team members to work in parallel. Once detailed dimensions and materials had been finalized, drawings were generated to guide the fabrication of the different mechanisms.

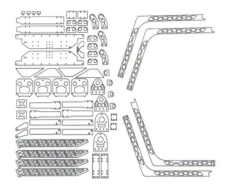

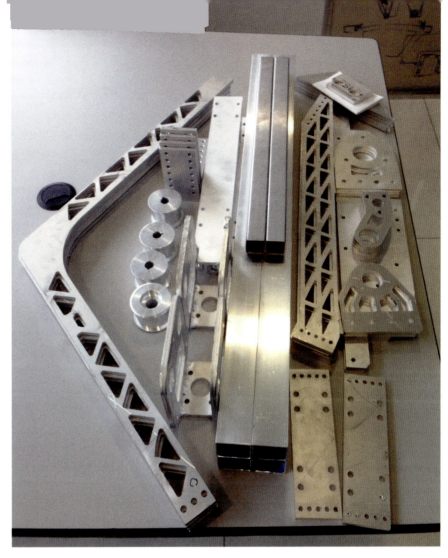

◆ ◗ *Certain three-dimensional modeled structural components were converted to two-dimensional drawings for laser and waterjet cutting at a local manufacturing company. Team members were invited to tour the facility and observe the cutting process. All other robot components were traditionally manufactured in the school's workshops.*

totes in the autonomous period, have an omnidirectional drive system, receive totes from the loading station, build a six-stack of totes, place a tote at any level in an existing stack, and raise a recycling container high enough to cap a six-stack of totes. Once all ideas had been reviewed, decisions were made as to which solutions would move on to the prototype phase. Prototypes were fabricated from recycled materials and previous robot parts found in the team's workshop. The major focus areas for development were the drivetrain and mechanisms to maneuver the totes and recycling containers.

MODULAR CAD AND PLAYING FIELD SIMULATION

The prototypes were refined until a final robot design was established. Then the students began to model the components using Autodesk Inventor software as a computer-aided design (CAD) tool. Robot dimensions were adjusted to fit within size restrictions and the prototype ideas were modified for seamless integration — all in a virtual environment. A modular approach to the design allowed different groups to work in parallel. The completed robot was then positioned in a virtual simulation of the playing field to validate that it interfaced with the field as intended. Once the model was complete, drawings were generated to manufacture the different components. By designing the entire robot in CAD, the team was able to identify materials, estimate weights, adjust dimensions, test modifications, and identify potential problems before fabrication.

TRAINING PREPARES STUDENTS FOR BUILD SEASON

LamBot used traditional manufacturing methods in its workshop to build the robot and took advantage of materials provided in the *FIRST*-supplied kit

○ *An omnidirectional drivetrain utilized mecanum wheels individually powered by a motor and gearbox. The aluminum chassis incorporated a laser cutout of the team number for added style and reduced weight.*

of parts when it could. The basic manufacturing operations at the school occurred in two different workshops. The primary workshop was used by the whole team for the majority of robot manufacturing. This workshop was equipped with hand and power tools, including a 16-inch scroll saw, a 14-inch abrasive cut-off machine, and a 17-inch drill press. In the months leading up to the season's start, the students were trained in proper power tool use and given the opportunity to practice their building skills with simple projects. Shop safety procedures were also reviewed, ensuring that every student was prepared to use these tools when the time came to build the robot. The second workshop at the school housed milling machines, welding machines, and lathes. Access to this workshop was limited to certain students and tool use required mentor supervision.

The team also utilized a local manufacturing company, Centro de Corte Potosino, for laser and waterjet cutting. The students designed the robot parts and components using Autodesk Inventor and SolidWorks software and then converted the designs into two-dimensional (2D) drawings that were emailed to the company. While the parts were being cut, team members were invited to the facility to observe the process and learn about the different machines and how they were used.

Once all individual robot components were manufactured, assembly was straightforward. The modular design of the robot allowed separate assembly of the drivetrain and the manipulators for the totes and recycling containers before being combined into a complete robot.

STAINLESS STEEL AND SLIDING RAILS

The robot's omnidirectional drivetrain featured four six-inch mecanum wheels, each powered by a MiniCIM motor with an AndyMark Toughbox Mini gearbox. The aluminum chassis was designed by students in Autodesk Inventor and sent to the manufacturing company to be laser cut. The chassis featured a cutout of the team number, a detail added for both aesthetics and weight reduction.

A tote manipulator stacked and delivered up to six totes to the scoring platform. The frame upon which the tote arms traveled was made from two parallel THK linear motion guides that each consisted of a rail and attached block sliding on ball bearings. The blocks on each rail were connected by a stainless steel plate to which the tote arms were mounted. Stainless

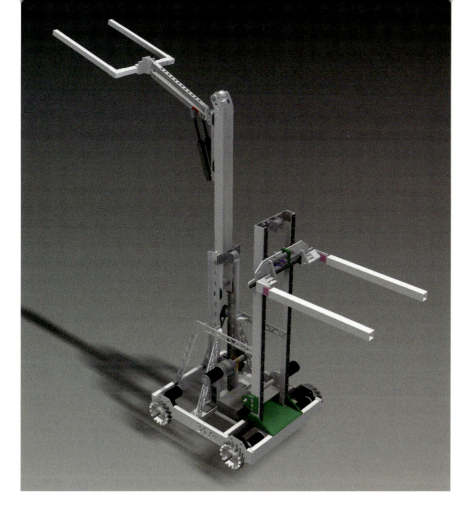

○ Performance at the robot's first competition revealed that the recycling container gripper design did not provide enough stability to prevent containers from falling off of tote stacks.

○ The initial robot configuration consisted of separate mechanisms for totes and recycling containers, located on opposite ends of the robot. Arms for collecting and stacking totes were raised and lowered with a pulley system, and a recycling container claw utilized a telescoping lifter with four-bar linkage.

steel was used for components that would bear the most weight from the tote stacks. The steel plate was raised and lowered with a pulley system. One end of a strap was connected to the steel plate, wrapped over a pulley mounted at the top of the frame, then extended down to the chassis where it was wound around a spool driven by a CIM motor with a three-stage AndyMark GEM500 gearbox. Mounted on the plate, in between the arms, were two opposing pneumatic cylinders. These cylinders actuated the opening and closing of the arms, which rode on a horizontal linear motion guide, to securely grab the totes. The arms were made of square aluminum tubing and had four tabs to secure the totes. The arms were opened and lowered over a tote, then closed, and as the system was raised, the tabs caught the lip of the tote to lift it.

To improve the accuracy of collecting totes, a roller mechanism was added after the team's first regional competition. Two square aluminum tubing arms were positioned below the tote manipulator arms, attached to the robot frame, and actuated by small pneumatic cylinders to open and close. The center of each arm was also connected to the chassis with surgical tubing to help them close faster. Mounted to the far end of each arm was a six-inch AndyMark rubber-treaded wheel connected to a BAG motor with a VEXpro VersaPlanetary Gearbox. The addition of the roller mechanism helped the robot collect totes faster, with more precision, and from different angles.

REVERSING A ROBOT COMPONENT FOR IMPROVED PERFORMANCE

Early on in the design phase the team knew it wanted a mechanism specialized for gripping the recycling containers. The mechanism designed to do this underwent several revisions throughout the build and competition seasons as the team integrated lessons learned. The initial version of the recycling container gripper consisted of a telescoping lifter and a claw connected by a four-bar linkage. The telescoping lifter was made similar to the one on the tote manipulator, with two parallel THK linear motion guides extended by a strap and pulley system with a CIM motor and GEM gearbox. The four-bar linkage was actuated by two pneumatic cylinders that elevated the claw at a consistent 20-degree

* The robot was reconfigured to stabilize the handling of recycling containers. The claw was rotated 180 degrees and shaped to match the profile of the containers. The four-bar linkage and powered lift were removed and the system used the tote manipulator for elevation.

* FRC Team 2478 proudly displays the team slogan on its shirts. Team members see victory beyond the robot, as they apply what they have learned to inspire others and influence Mexico's future.

angle to maintain recycling container stability. The lower linkage arm was constructed from stainless steel instead of aluminum for added strength. The claw was fabricated from lengths of square aluminum tubing, with an opening and closing motion actuated by a single pneumatic cylinder. The entire system was mounted on the side of the robot opposite to the tote manipulator. The gripper picked recycling containers up from the floor and elevated them high enough to place on top of a six-stack of totes.

The robot's performance at the team's first regional competition proved that it could stack and cap up to six totes, but the recycling container gripper was unstable and the container would often fall off of the tote stacks. The entire system was reconfigured to improve reliability and also reduce weight. The claw was rotated 180 degrees on the telescoping lifter so that it faced the same side of the robot as the tote manipulator. The gripper pulley system was removed, turning the telescoping lifter into a passive feature that simply guided the vertical movement of the claw. The four-bar linkage was replaced with a static piece of aluminum that held the claw parallel to the ground. The claw was also redesigned with a round profile that matched the curvature of the recycling containers, and an additional pneumatic cylinder was added for opening and closing. The final gripper design used the tote manipulator to grab and elevate recycling containers into the claw's grip. As totes were stacked below, they pushed the container and claw up, extending the telescoping lifter. When a fully built stack capped with a container was delivered to the scoring platform, the claw opened and, as the robot backed away, gravity lowered the telescoping lifter.

VICTORY GOES BEYOND WINNING

LamBot's slogan, "Victory goes beyond winning," was directly applicable to its experience in 2015. The team members were victorious in applying innovative and resourceful thinking to implement a successful redesign of the robot during the competition season. They confronted a challenge head-on and developed a solution that involved a 180-degree rotation of one of the major robot components.

The students' approach to the robot serves as a metaphor for a higher ambition, to initiate a 180-degree change in the mindsets of their peers who think "global problems are for adults to solve" and do not realize the power they have to make a difference in society. The team uses *FIRST* as a vehicle to inspire an interest in science and technology in Mexico's youth, and to promote to its community and country the importance of science and technology as the key to a better future. As for building robots, FRC Team 3478 believes that this is just an excuse to do something more important: to inspire others with the confidence to embrace challenge.

◯ With limited time and resources, FRC Team 3478 was able to successfully redesign a critical scoring mechanism on its robot. Creative problem solving led to rotation of one component and removal of unnecessary features, resulting in a simplified yet effective machine.

FIRST® IMPACT — TRANSFORMING A COMMUNITY

Lighting a Fire of Inspiration

The names emblazoned across the back of the Boyle Heights FIRST® Robotics Competition Team (FRC®) uniforms — purple hoodies — provide insight into the team's heritage. The names include Espinoza, Aquino, Bonilla, Flores, Sanchez, Estrada, Garcia, and others, with the number 4964 printed in large letters below each wearer's name. Boyle Heights in Los Angeles, CA, is sometimes referred to as the Ellis Island of the West Coast since it is home to a largely working class, immigrant population. Like many inner-city schools, Roosevelt High School in Boyle Heights struggles to provide opportunities for students who deserve much more than what the school can provide.

Boyle Heights is not exactly the place where one might expect to find a premier robotics team — but it is. Success in FIRST is independent of location. Instead, success in FIRST is dependent on the emotion, devotion, and dedication of the community. Such is the case for FRC Team 4964 — L.A. Streetbots — located in the Boyle Heights neighborhood of Los Angeles, CA.

An entry on the team's Facebook page celebrates its success with a post, "We, Team 4964, have gone against the grain of what of our community of Boyle Heights is. We did not merely build a robot; we built a legacy. We are not a community of drop-outs. We are the future." Similarly another post proclaims, "We have a great robot, but most importantly we have a great team." Such confidence, exuberance, and success did not just automatically happen. Rather, these accomplishments resulted from the efforts of a large team of supporters who rallied around a group of Boyle Heights students.

FIRST Team 4964 was founded in 2014 by long-time FIRST supporter, entrepreneur, and musician will.i.am to provide the youth in the very neighborhood that he grew up in with an opportunity they would not otherwise have. The team was sponsored by will and his i.am.angel Foundation as one of a number of initiatives to transform lives through education, inspiration, and opportunity. Dassault Systems, Raytheon, JP Morgan Chase, and other corporations provided additional support for the project. Individuals rose to the occasion to join the rookie team with employees from the sponsoring companies and other technology companies volunteering to mentor the L.A. Streetbots members.

Even with such high profile support, the team was not an automatic success for it faced many obstacles, including a lack of administrative and logistical support from the high school. This is where the team's street smarts kicked in, time and time again.

Faced with no room at the high school for a robotics workshop, the team converted a surplus trailer into a workspace. With no tools to fabricate the robot, families and businesses supplied hand tools and a small drill press. Lacking sophisticated manufacturing tools, the team compensated by making the most of the supplied robot kit components augmented with plywood and scrap aluminum.

Creative thinking was apparent throughout the team's work. As one example, a water bottle cap was repurposed as a crucial spacer in the drive system with the material providing the support, spacing, and wear characteristics needed for this application. Creativity was not limited to the robot itself, as a mentor provided motivation by allowing the team to shave his head the moment the robot first moved. Success came early when the team temporarily converted the workshop to a barbershop a bit sooner than the mentor expected.

On and off the field, FRC Team 4964 was — and continues to be — a success. The team finished strong in FIRST competitions and its accomplishments (and robot) were showcased with appearances on talk shows hosted by Queen Latifah and Arsenio Hall. Team members experienced opportunities well beyond the neighborhood of Boyle Heights, with tours of advanced manufacturing facilities, Space-X, and other high-tech companies, as well as attending the Los Angeles Women's Leadership Conference, now a regular part of what it means to be on FRC Team 4964.

Most significantly though, the students of FRC Team 4964 recognize that they, too, have a responsibility for making such opportunities available to others. In an amazing display of unselfish giving, the L.A. Streetbots students started a FIRST Tech Challenge (FTC®) team in their neighborhood middle school and have plans to start FIRST® LEGO® League (FLL®) teams in their local elementary schools. The FIRST experience at Boyle Heights is spreading like a wildfire.

Nearly one hundred years ago, William Yeats wrote, "Education is not the filling of a pail, but the lighting of a fire." Thanks to FIRST and the leadership of people like will.i.am, the neighborhood of Boyle Heights is a technology inferno that is transforming the lives of students who would otherwise not have such a precious gift.

▶ will.i.am's i.am.angel Foundation and others provide inspiration and support for the students on FRC Team 4964. The team's mentors and supporters provide opportunities that otherwise would not be available to this group of students.

CHAPTER 4
 THREE-DIMENSIONAL PRINTING

Additive manufacturing is a process where material is added during the fabrication cycle to create a physical object. Additive manufacturing is fundamentally different from traditional manufacturing where material is removed from base stock to create the final object. This process is commonly referred to as three-dimensional (3D) printing since layers of material are deposited on one another to build a designed component. 3D printing is a rapidly advancing industry that is becoming more accessible to a wider audience of users. The most common and least expensive 3D printing method in robotics is the extrusion of plastic material, a process known as fused deposition modeling. Photopolymerization is another common additive manufacturing method where layers of ultraviolet (UV) curable material are deposited on top of each other and subsequently cured with a UV light source. This chapter examines the use of 3D printing for robotics. The case studies in this chapter will provide examples of parts and entire systems that were 3D-printed and used in robotics applications. The case studies will illustrate how access to this technology promotes a new form of design thinking.

Three-Dimensional Printing | 99

Team 125 – Carefully Planned Incorporation of 3D-Printed Components

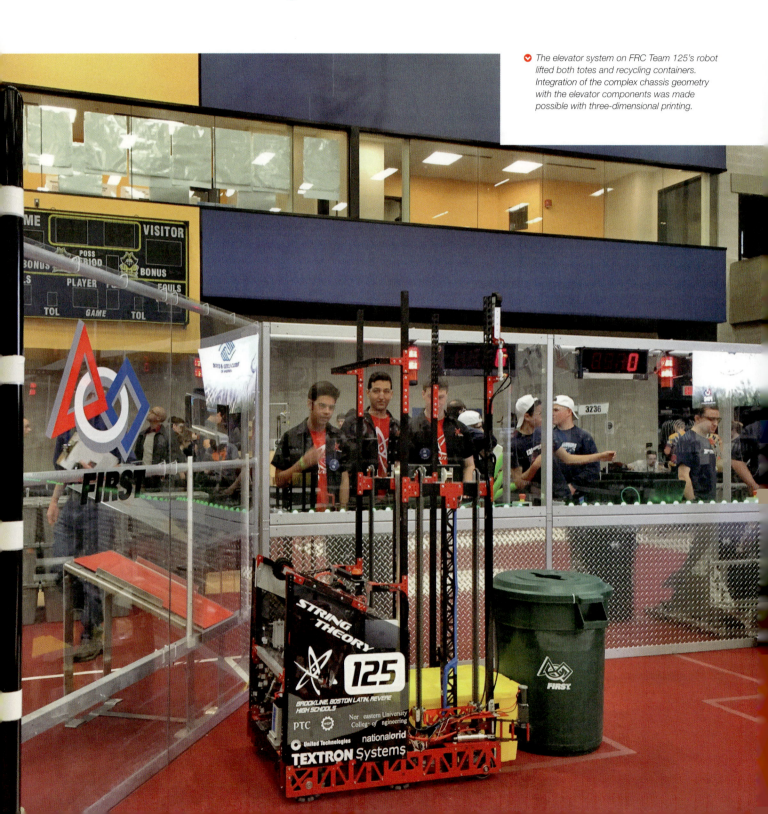

The elevator system on FRC Team 125's robot lifted both totes and recycling containers. Integration of the complex chassis geometry with the elevator components was made possible with three-dimensional printing.

The Stratasys Fortus 400mc 3D printer can print in four different layer thicknesses using 11 types of thermoplastics, which adds flexibility to the design process. This printer was used to produce the final components for the robot.

A SINGLE MOUNT SERVES MORE THAN ONE PURPOSE

The robot designed and built by FIRST® Robotics Competition (FRC®) Team 125, the NUTRONS, from Boston, MA, masterfully displayed intelligent application of three-dimensional (3D) printing. The flexibility to design complex shapes and structures, repeatable accuracy for creating duplicates, and speed of printing design iterations helped the team overcome limitations of traditional machining processes. Without this capability, the team wouldn't have been able to execute one of the main scoring features on its 2015 robot. While there wasn't a large quantity of 3D-printed parts on the robot, the quality and logical thought put into the sprocket mounts was an excellent example of design ingenuity.

ELEVATING TOTES AND RECYCLING BINS

The robot, named String Theory, featured a powered wheel intake that was pneumatically actuated to open and close. This mechanism was mounted to the front of the chassis for pulling totes and horizontal recycling containers into the robot from the playing field. Totes could also be loaded into the robot directly from the tote chute at the loading station through the back of the robot. The geometry of the back of the robot matched the slope and height of the tote chute, and as totes were loaded, the intake wheels on the front of the robot kept them from sliding through.

The robot's most unique feature was the elevator structure for lifting totes and recycling containers. Once the robot had collected a container through the intake in its U-shaped chassis, it backed up to the loading station. Two chain-driven parallel lifters were used to lift the recycling container as the first tote was loaded. Then the lifters were lowered, hooked under the tote lip, and raised again for another tote to be loaded through the chute. This was repeated until a capped six-stack of totes had been built in the robot, supported and stabilized by the sides of the elevator structure. Attached to the top of the front elevator uprights were two pneumatically actuated arms that aided in securing the recycling container during transport. When a capped six-stack of totes was delivered to the scoring platform, the recycling container was released and the robot backed away. These features, paired with a maneuverable

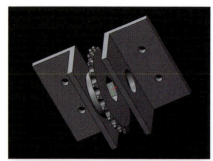
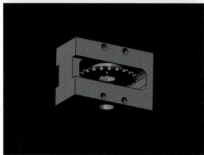

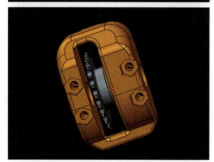

Design of the sprocket mounts evolved over four iterations to meet functional requirements. Computer-aided design software was used to model these iterations.

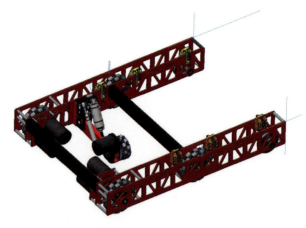

Computer-aided design software was used to model the robot chassis and sprocket mount placement to confirm proper integration prior to fabrication.

omnidirectional slide drivetrain and two arms to grab recycling containers from the step in the autonomous period, produced a very competitive robot.

SELECTING THE RIGHT PRINTER FOR THE JOB

FRC Team 125 used 3D printing for both prototyping and final fabrication of robot components. The team has a tabletop Cubify Cube 3D printer which has a build volume of 216 cubic inches. This printer uses filaments made from either recyclable acrylonitrile butadiene styrene (ABS) plastic or compostable polylactic acid (PLA) plastic to print in layers either 70 microns or 200 microns thick. This printer was primarily used for prototyping due to the printing resolution and material choices, which were not as desirable as the industry-grade 3D printers of the team's sponsors.

Final robot components were printed with a Stratasys Fortus 400mc 3D printer. This machine has a build envelop of 1,400 cubic inches and can print with 11 different thermoplastics, with four options for layer thickness, ranging from 127 microns to 330 microns.

With 3D printing capability, multiple iterations can be printed of the same part for rapid testing and redesign. The ability to create unlimited versions of a component to find optimal design is much more difficult with traditional machining methods, which can often require significant time and labor to generate each part.

The different robot components were designed with computer-aided design (CAD) software and then reviewed by the team's professional mentors before parts were fabricated. The NUTRONS used PTC Windchill to store all CAD files in a single location, from which team members could remotely access, edit, and collaborate on drawings. PTC, a *FIRST* Strategic Partner, donated this cloud-based product development and collaboration software to all FRC teams. The team's workspace is located on Northeastern University's campus, where it shares machinery with other clubs.

The sprocket mounts and drive shaft were carefully positioned to prevent interference with the chassis and drive wheels. Precise vertical alignment of the chains was essential for the elevator to function properly.

PRINTING A MULTIPURPOSE PART

When making the decision to incorporate an elevator in the robot design, the team acknowledged that it would have to be robust enough to handle the large game pieces, yet fast enough to be competitive at scoring points. A linear chain-driven system with mirrored lifters on each side of the elevator structure was the solution chosen to achieve the desired speed and required strength.

The system would require three sprockets mounted on a common shaft per side of the robot. Two sprockets would mesh with the vertical chain on each lifter, and the third sprocket would connect the shaft to the driving motor. The best method of mounting these sprockets was not initially evident due to how the elevator structure interfaced with the chassis and drivetrain. For the correct spacing of the elevator to handle the game pieces, the sprockets would have to cut through the inner structural side plates that sandwiched the drive wheels and chain. To overcome these complicated design concerns, the team decided to create specialized mounts for the sprockets.

FRC Team 125 knew that machining each mount would be time consuming and waste valuable material, so it chose to use 3D printing as the manufacturing method. The Fortus 400mc printer was used to make the final prints of the sprocket mounts due to its accuracy and capacity to print robust parts. Because 3D printing involves layering of material, the adhesion between the layers can be a concern, as forces applied to the component could cause them to break. The NUTRONS had access to higher-grade industrial printers, and so chose to use those with a stronger polycarbonate material to strengthen layer adhesion.

After further analysis and testing of the arrangement, additional design requirements for the custom mounts emerged. To hold the vertical chains in perfect alignment, the position of the sprockets and the shaft would have to be secured. To allow the shaft to spin freely while still being supported, the design incorporated bearings that were pressed into the mount.

The sprocket mounts were fastened to the side plates with socket head cap screws through mounting holes that were designed with hex patterns

◔ The NUTRONS received the South Florida Regional Chairman's Award, the most prestigious award at the event. This award honors the team that best represents a model for other teams to emulate and best embodies the purpose and goals of FIRST.

to capture the nut, thus eliminating the need for a separate wrench during assembly. Grooves were added to facilitate sliding the printed pieces into the plates, for easier mounting, and to better interface with the chassis.

3D PRINTING SELECTED OVER TRADITIONAL MACHINING

3D printing the sprocket mounts allowed the team to combine multiple features into a single structure. The mounts performed several functions, reducing the potential number of non-3D-printed parts that would be needed to achieve similar performance. The final design was also very complex, with intricate surface features to allow a secure fit within the drivetrain. These accents would have been difficult to machine using traditional manufacturing tools. Replication to create the six mounts and additional spares was also made easier with 3D printing.

Although much effort was put into sprocket mount development, incorporation into the robot helped to simplify the overall design. The unique mounts turned the elevator lifters into a single consolidated unit, rather than two components that independently worked together. The NUTRONS were a three-time winner of the 2015 Industrial Design Award sponsored by General Motors at two New England District Events and the New England *FIRST* District Championship. String Theory's strategic and effective incorporation of 3D-printed components was an integral part of the robot's sophisticated design and success on the field.

◔ The 3D-printed sprocket mounts incorporated hex patterns to capture the nuts on the socket head cap screws used to fasten the mounts to the chassis. Grooves were also incorporated in the design for a secure interface with the side plates.

◗ Team members assembled the sprocket mounts onto each side plate prior to constructing the full chassis. The mounts incorporated pressed-in bearings to allow the shafts to spin freely while still being supported.

Team 359 – Printing a Prototype

LEARNING BY DOING – ONE WHEEL AT A TIME

FIRST® Robotics Competition (FRC®) Team 359 knows quite a bit about building robots and building student skills. As the 2011 recipient of the *FIRST* Chairman's Award, the Hawaiian Kids Waialua Robotics Team realizes that the process of building a robot can take many forms, including designing and manufacturing custom parts that others might simply purchase from a vendor. Designing and manufacturing components exposes students to skills that can then be used to create other unique parts of the robot. The first step in the process is often prototyping an idea to ensure the correctness of a concept before it is manufactured. Three-dimensional (3D) printing is a technology that allows rapid prototyping to test concepts and confirm design specifications.

FRC Team 359 designed and constructed custom wheels for its robot for the 2015 competition season. While wheels were commercially available, the team favored designing and constructing its own wheels, with the tradeoffs in cost and time outweighed by the student learning experience. This creative manufacturing process progressed a hand sketch to a computer model that was used to first print the wheels in plastic. Once the design was verified the wheels were manufactured using aluminum. With eight wheels on the robot, the manufacturing process provided a host of opportunities for students to learn and practice the technologies involved in converting an idea into a functional object.

PROTOTYPING IN PLASTIC

The idea for the team's custom wheels originated as a hand sketch that was then modeled with computer-aided design (CAD) tools using Autodesk Inventor. This conversion — a translation of the hand drawing to an exact physical model described in computer code — was the first of a series of conversions of the idea from one format into another. The CAD model added specificity to the design with the resulting dimensions referenced continually through the ensuing manufacturing process.

Each wheel consisted of two sections. The front segment served as a plate that fastened to the load bearing back segment of the wheel. The back segment, which was closest to the robot's frame, included tapped holes to fasten the two segments together. The two sections sandwiched the wheel's tread and secured this tread to the hub. This approach of assembling the wheel from two separate sections facilitated the machining process used to create the hubs from aluminum stock. The robot design called for eight wheels, thereby requiring sixteen parts to be manufactured.

Before investing a significant amount of time manufacturing the wheels in aluminum, the design was first prototyped using 3D printing

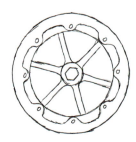

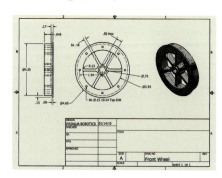

○ *A hand sketch captured the basic shape of the wheel, with definition added in the computer-aided design development of the initial concept.*

◐ The first layer of a 3D print of the wheels was created with tinted granular material that was chemically bonded to subsequent layers. Both sections of the wheels were printed at the same time.

technology to be sure the design was feasible and to ensure that the dimensions detailed in the CAD model were correct. The use of 3D printing accelerated this evaluation process and produced a full-sized model to evaluate the manufacturing and assembly process for the final design. As opposed to the many individual steps to remove material when manufacturing the actual wheels using aluminum, the 3D-printed prototype was produced in a single step, thereby enabling a quick validation of the design.

A Zprinter 450 3D printer was used to manufacture the plastic prototype of the wheel. This printer uses a process that binds individual grains of powdered plastic to create three-dimensional parts. Each of the segments was printed as individual cross sectional slices that were fused together. A powder base was first deposited across the bottom printer bed and the first layer of the objects was created using a tinted binding agent applied by the printing head at each point of the part.

A new 0.004-inch-thick layer of powder was then deposited over the previous layer, followed by the application of the binding agent. The depositing and binding process was then repeated layer by layer to build the part on itself. The completed product was post-processed with compressed air and soft brushes to remove residual powder on the part, with the residue recycled.

Parts printed in this fashion are very accurate but the printing process produces brittle parts. To eliminate the brittleness an additional binding agent was infused into the completed part to increase the component's strength. By doing so, the component was sufficiently hardened such that the back wheel segment could be tapped to fasten the two sections together. The result of this rapid prototyping was a full-sized replica of the design that verified the fastener locations and the alignment of the sections. Having ensured the correctness of the design, the actual wheels could be manufactured using aluminum.

◐ Infusing glue into the structure hardened the wheels and allowed the plastic to be machined and tapped like other common materials used to fabricate robot parts.

The blanks from the waterjet cutter were machined to reduce weight by removing excess material. CNC milling automated this step of the manufacturing process.

The computer-aided manufacturing instructions for each section of the wheel provided a series of commands to fabricate the parts on a waterjet cutter.

MANUFACTURING IN METAL

While the prototyping process relied on adding material one layer at a time to construct the component — a process known as additive manufacturing — creating the component in metal required that material be removed from a solid block of aluminum in a series of distinct steps. The manufacturing process began with cutting out the front and back pieces of each wheel using a waterjet cutter, a procedure that produced shapes with a uniform thickness. To remove material along the axis of the parts and to obtain the tolerances for the hub diameters, the waterjet-cut parts were machined on a lathe, a vertical knee mill, and a computer numerical control (CNC) mill to produce the final components with the specified dimensions. Each machining step required securely fastening the component in a piece of machinery and exactly positioning the cutting tool relative to the part — a time consuming process that amplified the need to ensure the parts were properly designed.

The first post-processing of the waterjet-cut parts involved placing the back wheel segment in a rotary table on a milling machine to cut away a 1.25-inch-wide and 0.3-inch-deep section of excess material from each wheel spoke. Both sections were then placed on the lathe, faced to the correct thickness, and cut to the designed diameter. While mounted on the lathe, grooves were cut in the lips of each segment to hold the wheel tread in place. For the back segment, the final manufacturing step on the lathe was boring a 0.5-inch diameter hole to ultimately support the drive shaft.

Holes were drilled using an indexing tool on the mill to create eight equally spaced holes in each segment. These holes were used to fasten the segments to a jig that secured the part in a CNC mill. A program on the CNC mill was run to cut away excess material on each wheel segment. To distribute the fastening loads a wide area was left around each fastener. This step significantly reduced the weight of each wheel. A drill press was used to enlarge the holes to fasten the wheel parts together, with the holes in the back segment tapped to retain the fasteners that held the two segments together.

An arbor press and broach converted the 0.5-inch round center hole of the back segment into a 0.5-inch hex hole to power each wheel with a hexed-shape drive shaft. The two segments for each wheel were then placed in a tumbler to deburr the edges. After wrapping the tread around the front wheel segment, the back wheel plate was aligned with the front plate and the fasteners were inserted to hold the two parts and the tread together. The result was a custom-designed, high friction drive wheel.

A SERIES OF TRANSLATIONS

The design and manufacturing process relied on a series of translations to manufacture the components. The first translation converted the original hand drawing into a CAD file that could be accurately dimensioned. This solid model became files formatted in a specific fashion, namely as a drawing (DWG) file, the native format of Autodesk Inventor, to convey the 3D parameters of the parts. The DWG file for each segment was converted into the stereolithography tessellation language (STL) format to create a set of layer-by-layer instructions that specified where the 3D printer would deposit material and binding agent. Just as the CAD file was a representation of the original hand drawing, the STL file was an approximation of the solid model since the STL format is based on the assembly of triangles to create a faceted model of the 3D shape.

The CAD software also produced a drawing exchange format (DXF) translation of the original DWG file. This format is commonly used to exchange files between software applications. A proprietary software converter associated with the OMAX waterjet used by the team created yet another version of the original file in an OMAX Routed Data (ORD) format that produced the tool path to position the high-pressure water nozzle and cut each part. Similarly, Mastercam software was used to create the G-code — language that directs numerical controlled machines to fabricate parts. G-code contains embedded instructions on the location,

Once machined, the front and back hubs were tumbled to remove manufacturing burrs. Tread was sandwiched between the two sections that were fastened together, with the compressive force between the sections holding the tread in place.

○ *Four wheels on each side of the robot provided multiple points of contact for the drive system, even when climbing the scoring platform. The tread's high coefficient of friction ensured a secure grip on the carpet and platform.*

speed, and pathways for cutting tools in computerized machines. As illustrated in this example, specialized computer-aided manufacturing (CAM) software served as the translator and interpreter for each file format. Advances in these software packages have increased the ease of using different tools to manufacture individual components.

RAPID PROTOTYPING AS A DESIGN TOOL
Using rapid prototyping as a preliminary tool to validate design concepts is an important step when many parts need to be manufactured. For FRC Team 359, the process of designing and manufacturing their own wheels was an important training opportunity to teach students about computer-aided manufacturing processes.

With sixteen separate parts needed for this robot component, the process provided a training opportunity for new students to operate the machinery needed for each step of the manufacturing process. More experienced students — some having been trained during the production runs of the first sets of wheels — provided oversight to the new students. Many wheels with multiple parts and many steps to manufacture each part resulted in a method to not only manufacture the wheels, but also to train team members on design and manufacturing techniques.

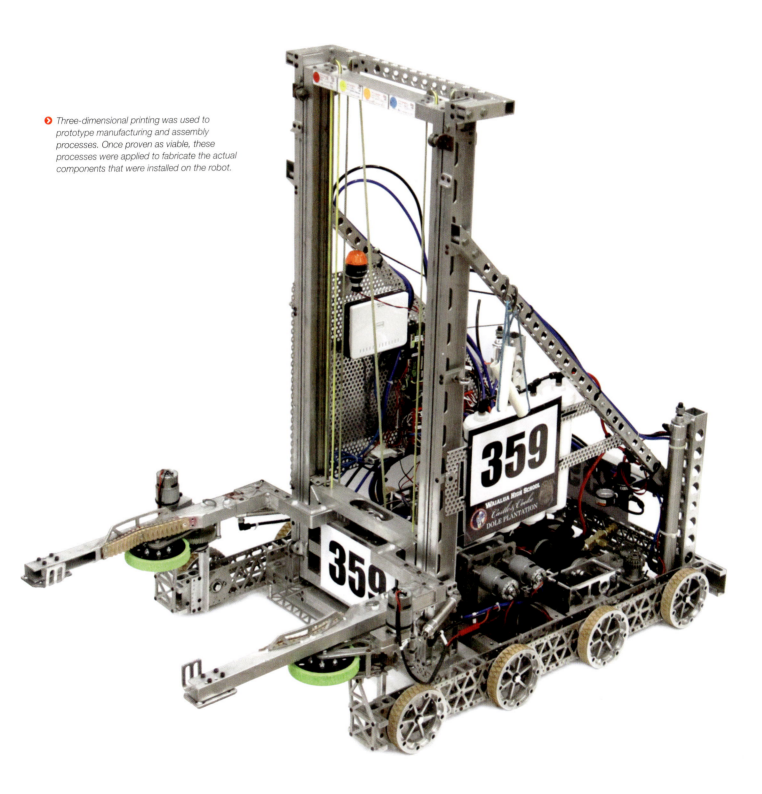

◐ *Three-dimensional printing was used to prototype manufacturing and assembly processes. Once proven as viable, these processes were applied to fabricate the actual components that were installed on the robot.*

Team 2601 – Additive Manufacturing Inspiration

Three-dimensional-printed parts were used in the drivetrain and active tote intake system to manipulate game pieces on the field. Additive manufacturing was used in place of custom machining to fabricate critical components.

3D PRINTING INSPIRES CREATIVE SOLUTIONS

Rapid prototyping and manufacturing were the foundation of *FIRST*® Robotics Competition (FRC®) Team 2601's design and build process. The Steel Hawks, from Flushing, NY, used innovative design and three-dimensional (3D) printing to conquer the manufacturing challenge of building a robot with limited machining resources. The team members' knowledge and skill in additive manufacturing inspired unique design solutions on which they could quickly iterate to continuously improve performance.

FRC Team 2601's robot, Raptor, was designed to manipulate totes and recycling containers from the field. The machine's design relied heavily on 3D printing capability, and evidence of this could be seen in almost all major components. Along with creating static components, they also used 3D printing to develop interfaces between components that would normally require custom machining, such as between gearboxes and wheels from different suppliers. Printed components were analyzed and tested to find the appropriate balance of production speed, design complexity, weight, strength, and reliability for the particular application. When this manufacturing method couldn't be applied, team members made smart decisions to accommodate their limited manufacturing resources,

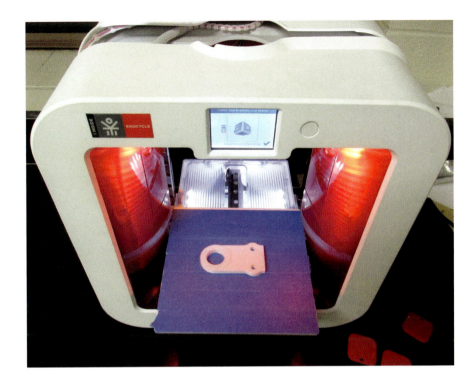

> The team's EKOCYCLE Cube 3D printer was donated by 3D Systems and Coca-Cola. This printer's filament cartridges use post-consumer recycled material.

using commercial off-the-shelf components and an existing inventory of hardware when possible.

Raptor collected game pieces and created stacks with three simple and robust subsystems: a drivetrain, intake, and elevator. The unique drivetrain and active intake were made possible by the extensive use of 3D-printed parts. The AndyMark AM14U chassis provided in the kit of parts was used as the platform for the omnidirectional drivetrain. Six wheels were powered by four CIM motors with AndyMark Toughbox Mini gearboxes. A center wheel, powered by a single CIM motor and Toughbox Nano gearbox, enabled side-to-side mobility.

An active intake system was included to draw totes and containers into the robot. Arm-mounted wheels were powered by AndyMark 9015 motors and planetary gearboxes. Polycord and pulleys drove the wheels, allowing some slip as to not wear out the motors and wheels. The arms opened and closed with two-inch stroke pneumatic cylinders. A chain-driven elevator and hooks then lifted the totes and recycling containers to build and cap stacks.

MULTIPLE 3D PRINTERS TO MEET DEMAND

The Steel Hawks built Raptor despite having extremely limited access to machining resources. The only stationary power tools used in the manufacturing process were a drill press and table saw. Rather than perceive this as a roadblock, the team took the opportunity to redefine how a robot could be built. To supplement the areas where metal machining wasn't an option, FRC Team 2601 turned to 3D printing. Three different 3D printers were used in the fabrication process to take advantage of the unique printing attributes for different printed pieces, and to meet the demand for quantity of parts produced.

The team applied for and received a free EKOCYCLE Cube 3D printer donated by 3D Systems and Coca-Cola. The filament cartridges used by this 3D printer each contain post-consumer recycled material equivalent to three recycled 20-ounce polyethylene terephthalate (PET) plastic bottles. The EKOCYCLE Cube uses plastic jet printing (PJP) technology, a thermoplastic extrusion process very similar to fused deposition modeling (FDM). It has a build volume of 216 cubic inches and a layer resolution of 200 microns. This printer was best for parts that needed to be printed fast, and when the smoothness of the surface finish wasn't a requirement.

The team received a grant to purchase a MakerBot Replicator 2 3D printer for the 2015 season. This FDM printer has a build volume of 10 cubic inches and can print with a 100-micron layer resolution. This printer was slower than the EKOCYCLE Cube,

◔ The final iteration of the suspension system used 3D-printed brackets to mount the CIM motor, which drove the center wheel. The wheel was suspended with springs that connected the gearbox to the chassis.

but resulted in printed parts with a smooth surface finish that didn't require sanding. It printed with a polylactic acid (PLA) filament, a biodegradable thermoplastic derived from corn.

The team also purchased a MakerBot Replicator Fifth Generation 3D printer, which had the same print technology, filament, and print resolution as the MakerBot Replicator 2. However the 456-cubic-inch build volume was 11% larger than the team's other MakerBot printer and it included an onboard camera for monitoring print progress. Once the robot components had been modeled in software, they could be printed on any of these machines.

AN AGILE DRIVETRAIN MADE POSSIBLE BY 3D PRINTING

Upon evaluation of the game, the team decided to create an agile drivetrain that had more than just the forward/backward movement of a traditional tank drive. The team selected a slide drive configuration, also referred to as an H-drive, which is similar to a tank drive but with an additional wheel perpendicular to the rest. This gave the robot the ability to turn on its own axis and translate sideways. Six four-inch VEXpro Omni Wheels were mounted on the chassis, three per side, in a traditional tank drive format. Centered between these and shifted slightly to the back of the robot was a larger six-inch omnidirectional wheel, rotated 90 degrees to the rest of the drivetrain to provide the lateral movement. When tested, the system performed well on flat surfaces but was not ideal on an uneven floor or carpet. To remedy this, the team designed a suspension system to keep the center wheel in contact with the floor.

The first iteration of the suspension system used a belt system to transfer power from the CIM motor and driveshaft, with 3D-printed tensioners to keep the belt taut. The large weight of this system and mechanical complexity led the team to seek another solution.

The second iteration utilized 3D-printed brackets made with the MakerBot Replicator 2 to hold the CIM motor. The 3D-printed brackets experienced high stress loads and so were continuously printed to save as spare parts in the event of failure. These spares were made using multiple printers simultaneously to meet production goals. The relative ease of developing lightweight parts with the 3D printers was critical to the team creating a successful drivetrain.

INTAKE EVOLUTION TAKES ADVANTAGE OF RAPID PROTOTYPING

To facilitate the collection of totes and recycling containers and reduce cycle time, the Steel Hawks developed an active intake device. The design had to take into consideration the irregular and challenging shapes of the game pieces. A variety of different power transfer and actuation devices were prototyped to discover the best solution. Bearing brackets, pulleys, hubs, and motor mounts for these prototypes were all 3D printed. Three iterations led to the intake mechanism installed on the robot the last day of the build period, but iteration continued into the competition season.

The initial prototype used gearmotors with a 0.375-inch outer diameter shaft to directly drive BaneBots wheels with a thermoplastic rubber tread and a 0.75-inch inner diameter hex mount. Custom 3D-printed hubs were created to overcome these disparate dimensions. The system was not as fast at manipulating the game pieces as the team hoped, and the placement of the expensive motors put them at risk for damage.

A revised version was created to increase the speed and protect

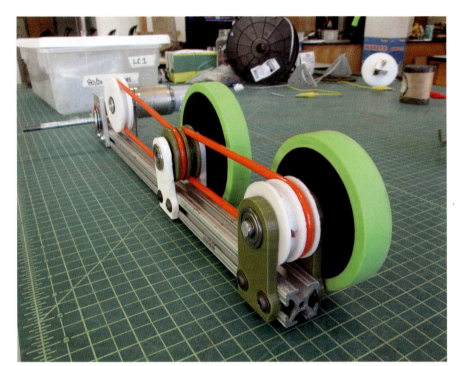
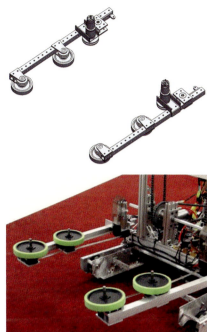

Three prototype iterations of the active intake system led to a design that included two sets of polycord-driven wheels. For the competition robot, the 3D-printed bearing brackets, gearmotor mounts, and clevis brackets developed for the prototype were replaced with aluminum equivalents.

the motors. The gearmotors were relocated and power was transferred to the intake wheels with polycord and pulleys, printed with the MakerBot Replicator Fifth Generation. The custom 3D-printed pulleys were designed with threaded set screw holes to mount to the steel shaft. The team also printed gearmotor mounting brackets and bearing brackets to hold the intake wheels with both the MakerBot Replicator 2 and the EKOCYCLE Cube. The system was an upgrade from the first design, but there was still room for improvement.

The third iteration of the active intake was smaller than the first two designs. It incorporated two sets of wheels to grasp the game pieces. Pneumatically actuated arms were added to help with alignment of the pieces and to hold them more securely. The system still used brackets, but the aggressive process of collecting totes from the landfill was causing them to structurally fail, despite a 100% infill of the PLA filament. This failure was addressed by a redesign to increase the thickness of the brackets and integrate them with aluminum for rigidity.

By the end of the build period, the system consisted of two arms that extended wheels 19 inches beyond the front of the robot that opened and closed by pneumatic cylinders. Power from the AndyMark 9015 motors and a 71:1 reduction planetary gearbox was transferred to the wheels with polycord, allowing for fast acquisition of totes and recycling containers. The 3D-printed bearing brackets, gearmotor mounts, and clevis brackets developed for the prototype were replaced with aluminum equivalents donated by sponsor Magellan Aerospace.

CONTINUED ITERATION FOR DIVERSIFIED STRATEGY

System iteration continued beyond the build season. The team had initially focused its strategy on rapidly collecting and stacking totes from the landfill zone, however as the season progressed the team saw that tote acquisition from the loading station was very effective. The intake system was revised to reduce weight and more effectively collect totes from the loading station. Two of the four intake wheels were removed after the modification underwent rapid proof of concept testing using 3D-printed brackets made on the EKOCYCLE Cube 3D printer. Once the effectiveness of the shortened intake system was demonstrated, the system was modified in time for the robot to compete at the team's second regional competition.

◐ Iteration of the intake system continued beyond the build season. Two of the four intake wheels were removed and the intake arms were shortened, with proof of concept testing achieved using the EKOCYCLE Cube 3D printer. These modifications added capability, enabling the robot to collect totes from the loading station.

The final intake design exhibited speed and reliability when acquiring totes and recycling containers.

The team's knowledge and application of additive manufacturing were crucial to the robot's success. Team members used this technology as the foundation for their design and build processes, with workflow based solidly on rapid prototyping and manufacturing.

Although it may seem like FRC Team 2601 had limited resources, its skill at 3D printing allowed the team to save time and money through the transition of idea to prototype, and prototype to final product.

◐ Three-dimensional printing was integral to FRC Team 2601's design and build processes, from rapid prototyping to final manufacturing. With limited access to machining resources, the team developed unique solutions through additive manufacturing.

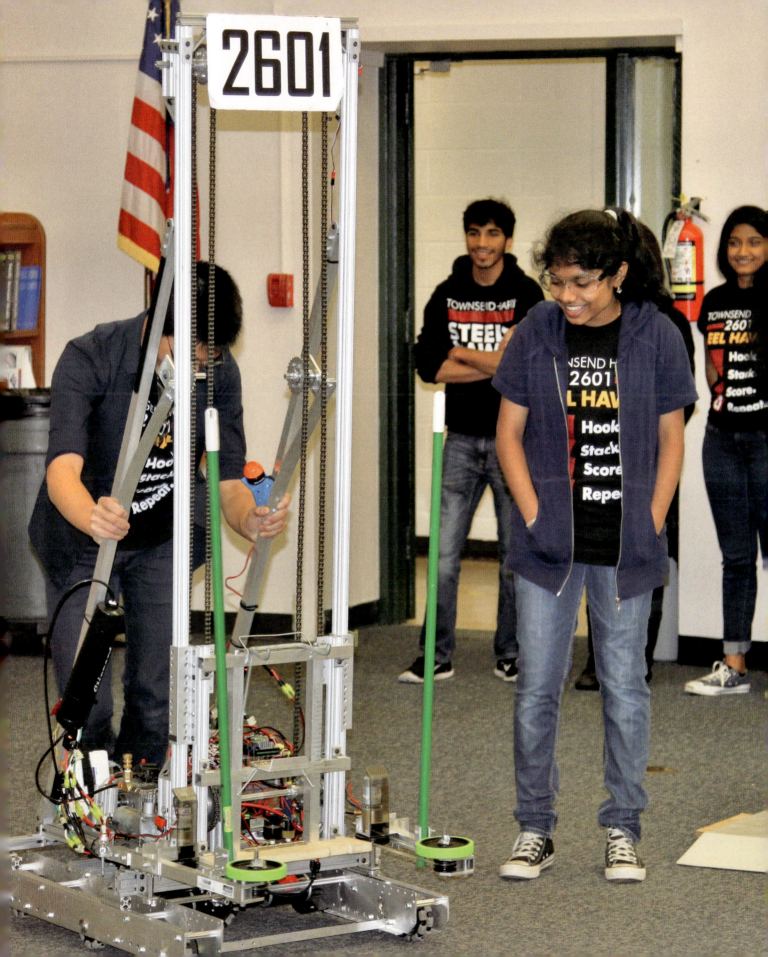

Team 3824 – Large Scale Additive Manufacturing of Robots

Three-dimensional printing was used to manufacture the majority of the robot's structural components. The printed parts were rapidly produced using 3D printing technology capable of fabricating more than 40 pounds of components in one hour.

The combination of CAD and 3D printing enabled a seamless transition from digital design to production. The digital model was converted into a series of commands that provided instructions for the advanced 3D printers at the Oak Ridge National Laboratory's Manufacturing Demonstration Facility.

MAKING THE MOST OF MANUFACTURING RESOURCES

When the newly found FIRST® Robotics Competition (FRC®) Team 3824, Hardin Valley Academy RoHAWKtics from Knoxville, TN, jumped into robotics in 2011, they faced a significant challenge. Though the team had motivated students and mentors, its newly constructed school did not have a machine shop. The school did however have a new Stratasys µPrint three-dimensional (3D) printer. As fortune would have it, the list of mentors included an innovative teacher and an expert in additive manufacturing technology. Faced with a motivated team and adversity, the pair conspired to convert the challenge into an opportunity. To combat the lack of a machine shop, the team planned to print as many robot parts as they could. The team immersed itself into learning how to use SolidWorks computer-aided design (CAD) software and create components such as mounting brackets and motor couplers using the school's 3D printer. The first year's learning curve was steep, but a regional competition finish as the top performing rookie team propelled FRC Team 3824 to the FIRST Championship, a season finale that has since been repeated many times.

In 2012, the team shifted its workshop from the high school to the nearby Manufacturing Demonstration Facility (MDF) at the Oak Ridge National Laboratory in Tennessee. The MDF houses state-of-the-art polymer, metal, and composite additive manufacturing capabilities to help the industry adopt economical, efficient, and sustainable manufacturing technologies. The team's move not only provided the students with access to cutting-edge manufacturing technology, but also opened the door to learning about the latest technology developments from some of the field's leading researchers. Starting that year and continuing every year since then the team has printed as many components as possible using additive manufacturing technology.

The partnership quickly advanced the team's awareness of this technology and even influenced the development of large scale 3D printing. It is no surprise that FRC Team 3824 has become known as "the team that prints robots."

ADDITIVE MANUFACTURING METHODS AND ROBOT COMPONENTS

Unique fabrication resources surrounded the team at the Manufacturing Demonstration Facility. Where commercial printers might be considered to be limited in their range of printing materials and slow in the speed of printing, those limits were generally eliminated for FRC Team 3824. Additive technology machines at the MDF printed with a variety of high tech material including carbon fiber-infused acrylonitrile butadiene styrene (ABS) and fiberglass-infused ABS. The fabrication speed was faster as well, with 3D printers capable of printing 40 pounds of material in an hour available for the team's use. Since the MDF

● Known as the "team that prints robots," FRC Team 3824 first used 3D printing to fabricate all of the structural elements on the team's robot in 2012.

● 3D printing was an efficient process to manufacture many specialty components on the 2015 FRC Team 3824 robot. Parts were manufactured with orientations that maximized each component's ability to withstand the expected loads.

● In 2013, essential 3D-printed components were wound with carbon fiber to increase the strength of the composite structures.

additive manufacturing machines used commercially available plastic pellets instead of the specially manufactured ABS filament, material costs were much lower than traditional filament printing.

Even without access to advanced 3D printers, additive manufacturing technology benefits the fabrication of small (less than one cubic foot in volume) complex parts such as pulleys, sensor mounts, and motor shaft adapters since machining small numbers of such components is very time intensive. A number of factors influence the decision to 3D print or traditionally manufacture any single component, including the ability to minimize the weight of the fabricated part; reduce the overall part count by combining features into a single design; use a sparse internal fill to reduce weight while maintaining structural integrity; and augment the printed material with hardware, such as threaded inserts, to interface the part with other components.

FRC Team 3824's first experiences using additive manufacturing for robot assembly centered on printing pulleys, motor couplings, and structural components. That work expanded in 2012 to creating mechanisms, motor mounts, and large bearings, as well as the chassis and all of the robot's supporting structure. Given the volume of parts that were produced using additive manufacturing technologies, the year marked the first time the team produced a "3D-printed robot."

The following year, carbon fiber winding methodologies were combined with additive manufacturing techniques to increase the strength of the fabricated components. Structures subject to high loads, such as the chassis, were strengthened by mounting the printed parts on a filament winding machine that covered the plastic structure with tightly wound strands of carbon fiber. This combination of 3D printing and carbon fiber winding created a composite structure that was very strong and extremely durable.

By 2014, the Manufacturing Demonstration Facility had perfected its Big Area Additive Manufacturing (BAAM) technology capable of quickly printing components with footprints

Big Area Additive Manufacturing (BAAM) technology quickly printed large components of the robot. Each of the ramps to load totes was printed as a single part that was bolted to the robot's frame.

as large as eight feet by eight feet — a scale ten times that normally associated with 3D printing. The unit originally could print at a speed of ten pounds of material per hour, with this speed eventually increasing to 40 pounds of material per hour. Plastic pellets with a diameter of 0.3 inches served as the raw stock that was heated and extruded by the BAAM printers. By using commercially available plastic feedstock, costing between $2 and $5 per pound, the material expenses plummeted well below the standard cost of $30 per pound of ABS filament.

The BAAM machine was used to print large structural elements with conventional additive manufacturing printers used for parts requiring finer detail. The speed of BAAM printing was impressive, with the entire robot chassis printed in less than two hours. This experience set the stage for the team to apply even newer additive manufacturing technology during the 2015 FRC season.

THE 2015 BAAM-BOT

Whereas the previous year's advancement for FRC Team 3824 was the use of large area additive manufacturing machines, the team's technology advantage in 2015 was infusing additional materials into the plastic beads that fed the BAAM printer. With this development in additive manufacturing materials, FRC Team 3824 printed its 2015 robot components using two types of composite materials. Carbon fiber-infused ABS was used to print the chassis to take advantage of this composite's high strength and inherent stiffness. For BAAM-printed parts that needed flexibility, such as the arms to grasp containers, fiberglass-infused ABS was the material of choice. The cost to print the robot was approximately $100 due to the low cost of the pellet feedstock. Twenty-five pounds of parts for the chassis, forklift, and truss were printed in only two hours to create FRC Team 3824's BAAM-Bot.

A printed pair of sleds guided the totes from the loading station into the elevator on the far side of the robot. The sleds were originally manufactured on the BAAM printer using carbon fiber-infused material and later reprinted using fiberglass-infused material to improve the durability of the appendages. Given the substantial thickness of the BAAM-printed parts and the wide cross sectional area of the printed material, the finished frame offered some of the same advantages as traditional robot structural material for adding components. Similar to fastening techniques used on aluminum frames, the BAAM components could be drilled to attach components directly on the 3D-printed frame members.

▲ The technology and processes to print robot parts were modified by the team and the lab's research staff to produce a printed car and a printed auto body as advanced technology demonstration projects.

Other robot components, such as motor mounts, shaft adapters, pulleys, and fingers to hold suspended totes were printed using smaller scale additive manufacturing technologies. Embedded in the team's experience of their prolific use of 3D-printed parts was the refinement of CAD skills for members of FRC Team 3824. The ability of individual team members to use SolidWorks software grew with each additional part that needed to be printed. A design was first created as a software model and then exported to the printers for manufacturing, thereby establishing a fast path from design to product realization. This combination of design skills and innovative uses of technology served FRC Team 3824 well during the 2015 season where they won two regional competitions and earned an invitation to the FIRST Championship.

ADVANCING ADDITIVE MANUFACTURING TECHNOLOGIES

In addition to building and applying developing technology to fabricate robots, FRC Team 3824 members also had a unique chance to advance additive manufacturing technology in partnership with one of their sponsors. Research opportunities were available for team members at the Oak Ridge National Laboratory's Manufacturing Demonstration Facility where students worked on novel robotic drive systems that took advantage of the advancing technologies in additive manufacturing. Their work fabricating complex mechanical drive systems using additive manufacturing technologies was presented at a number of engineering forums and eventually patented by a group of students.

FRC Team 3824 alumni were also recruited to develop software methodologies for modifying large scale CAD files to optimize 3D printing. Their research addressed efficient algorithms to manipulate the large models and define the print specifications, including supports for overhangs, arranging polygonal shapes to optimize void filling, and modifying print paths to increase material adhesion. This work improved the efficiency and effectiveness of Big Area Additive Manufacturing technology that was demonstrated in a partnership with Local Motors to create the Strati, the world's first 3D-printed car.

Buoyed by this success, the lab challenged itself to a six-week FRC build season timeline to design, print, and assemble another demonstration project using the most advanced additive manufacturing technologies. Using the lab's newly developed techniques for printing, finishing, and painting the 3D-printed car, a replica of the original 1965 Shelby Cobra was completed in six weeks and showcased at Detroit's 2015 North American International Auto Show. This stop was merely a warm-up act as the car's premier display was at the 2015 FRC Smoky Mountains Regional where it was paired with its family relative, the 3D-printed robot created by FRC Team 3824.

The partnership between the students, mentors, and sponsors not only produced a collection of vehicles fabricated using additive manufacturing technologies, but also developed new knowledge, expertise, and innovations that advanced the additive manufacturing industry. With an ultimate goal of printing a robot within one hour, FRC Team 3824 is rapidly shifting the landscape of robot design and rapid manufacturing.

Materials were combined in the printing process to improve performance. The strength of the robot's base was increased using a combination of plastic and carbon fiber in the printing process. Glass fiber was added to the printing process to improve the flexibility of the container-grabbing arms.

Team 5030 – Fast Solutions for Small Parts

◆ Table-top 3D printing equipment enabled specialty components to be rapidly manufactured for a FIRST robot, including many of the bearing and electronics mounts, spacers, brackets, and sensor triggers.

PRINTING INSTEAD OF MACHINING

Additive manufacturing technologies have many applications on *FIRST*® Robotics Competition (FRC®) robots, ranging from printing robot frames to fabricating small components. At the smaller end of this range, three-dimensional (3D) printing is often the most efficient manufacturing process. In many FRC applications, mounts, spacers, and simple devices for FRC robots can be quickly designed and manufactured using 3D printers. The use of this technology also has another potential advantage: making the best use of limited time. By initiating print processes at the end of a team's work period, the manufactured part can often be ready at the start of the next work period. Though team members need a break, manufacturing processes can continue around the clock.

FRC Team 5030, The Second Mouse, from New Hartford, NY, benefitted from initiatives within *FIRST* to acquire two 3D printers: a Cube 3D printer and an EKOCYCLE Cube 3D printer. Both machines produce similarly sized objects, with the EKOCYCLE Cube's six-inch by six-inch print area slightly larger than the 5.5-inch by 5.5-inch print area of the Cube. While the Cube 3D printer can use acrylonitrile butadiene styrene (ABS) and polylactic acid (PLA) plastic filament, the EKOCYCLE Cube 3D printer uses filament created from recycled plastic. With limited access to other manufacturing equipment, 3D printing evolved as FRC Team 5030's go-to method for manufacturing small robot components.

MANUFACTURING INNOVATIONS

FRC Team 5030 designed and printed a collection of creative parts for its robot. Limit switch mounts securely held the sensor and allowed easy adjustment when exactly positioned on the robot's aluminum elevator supports. Another sensor application was a device mounted on a rotating shaft that tripped a limit switch once per revolution. The trigger and limit switch combined to create a simple and inexpensive encoder to measure the elevator's travel distance based on the rotation of a lead screw.

In other applications, 3D-printed parts provided structural integrity for robot components. Spacers were fabricated to serve as bolt guides within the channeled robot frame where they prevented the frame from being crushed when the bolts were tightened. Other 3D-printed spacers were used to best position the elevator hooks. By modifying the amount of fill used to print the spacers, the compression of the devices was modulated to provide the ideal amount of compression to securely grip the acquired totes. The resiliency of compressed plastic was also taken advantage of in the design of a protective collar on the elevator lead screw. The plastic collar prevented the load-carrying nut on the lead screw from slamming into the screw bearing. The contact shock was mitigated when transferred to the shock-absorbing plastic collar.

The electronics system also benefitted from the team's familiarity using

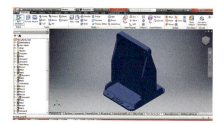

⬤ *Custom pieces, such as a tower to vertically stack speed controllers, were designed in CAD and printed to efficiently utilize the robot's limited real estate. The printed parts included mounting holes for the electronics and the bracket.*

3D-printed material. Small clips were fashioned to retain power and signal wires within the channels of the robot's structural aluminum frame. The channels kept the wires free from entanglement and protected the wires from damage. A tower to efficiently stack four speed controllers in a compact space was also fabricated on the team's 3D printers. The tower was designed with runways for the wires and mounting holes for the speed controllers.

An iterative process was used to design and manufacture bearing blocks for the robot's drive system. The printed bearing blocks were designed with the mounting locations precisely aligned with pre-drilled holes in the frame members. Multiple designs were created and tested to be sure the 3D prints had the strength to withstand the driving system loads they were subjected to. Because of the limited structural strength of plastic, aluminum supports were fabricated for the front bearings that experienced higher loads.

PRINTING FOR OTHER PURPOSES

These parts were fabricated following a mockup stage when each part was conceived. Aided by the physical mockup model, the part progressed to a digital model using computer-aided design (CAD) software. The resulting files of the CAD models, as well as the 3D printer command files, were carefully managed to keep the production process organized. The process to produce these parts also included a testing cycle to ensure the components were correctly sized and had the required strength before they were installed on the robot.

The amount of plastic fill within each part's voids was modified to support the load on each part. Once the part's geometry and strength were verified, with improvements made as needed, multiple copies of the part were manufactured, with the extra parts used as spares. As a biodegradable thermoplastic, PLA was the preferred printing material. With this characteristic any incorrect components were recycled.

The team made use of the ability of the 3D printers to operate automatically because designs were often loaded into the printers at the end of a work

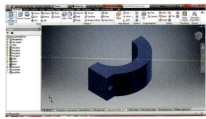
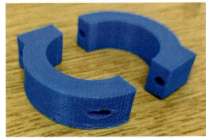

◐ ◑ *Shaft collars were printed and mounted to the bottom of lead screws to prevent the drive nut from colliding with the less-forgiving steel motor bearing.*

period, printed while the team was absent from the space, and available for use the next work period. This ability to manufacture during off-hours benefitted the team's productivity and allowed one idea to be tested while a second idea was being developed. In addition to the time saved when 3D printing, the parts also saved weight when compared to metal alternatives.

In addition to being a quick method to fabricate custom parts, FRC Team 5030 also used 3D printing methods to expose students to new technology. The ability to design an object on a computer and then create that object as a physical device was a captivating tool to engage students in the design process. This approach was used not only to recruit new team members and have those members instantly contribute to the team, but also as a means for sharing technology with younger students during outreach programs. A clever extension of this approach to pique interest in new students was printing miniature versions of FRC game pieces as a means to share the team's history participating in *FIRST*.

WISELY USING RESOURCES

3D printing allowed FRC Team 5030 members the chance to be creative as they fashioned solutions to mounting, spacing, and loading challenges on their robot. Attention to accuracy and precision, similar to the skills needed during metal machining operations, was required to fabricate parts that fit and functioned properly. Given the team's use of 3D printing technology to align and retain the robot's drive bearings, high levels of precision were required to ensure that the drive system worked during every match.

As a two-year-old team founded in a dorm room, FRC Team 5030 had limited access to the traditional and advanced manufacturing equipment that other robotics teams often have access to. By making the most of the available resources, the team relied on 3D printing technology to create high precision parts that increased the team's productivity and the robot's performance. Along the way, the team demonstrated creativity while using new technology, and tenacity while resolving challenging problems.

◯ Blocks for the drive system's bearings were modeled and manufactured using 3D printing technology. The close tolerances of 3D printing allowed the metal bearings to be simply press fit into the plastic parts.

◯ The availability of 3D printing technologies provided FRC Team 5030 with an ability to easily fabricate specialty components that they otherwise would not have been able to manufacture.

◯ The elastic properties of plastic were exploited in the design of a shaft collar to absorb the load of a rapidly driven lead screw, preventing the slamming force from damaging the bearing, transmission, and motor.

◯ Printed parts included devices for a rotating shaft to trigger a limit switch and count rotations, shaft spacers, and limit switch mounting brackets.

FIRST® DEAN'S LIST

Students Share their FIRST® Experience

Inspiration and recognition are the heart of the FIRST® experience. On one level, inspiration and recognition are obtained through the process of designing and building a robot, where students and mentors are inspired by the teamwork, comradery, and success of seeing their ideas come into being, and recognizing that they have the power to create. The competitions present another opportunity for inspiration and recognition when individuals see the work of other teams. This inspiration and recognition is not only based on the robots, but more importantly, on the people who make up each FIRST team.

The FIRST Dean's List Award celebrates FIRST students and recognizes these student leaders who motivate their teams to fulfill the mission of FIRST and effect culture change. The FIRST Dean's List Award also celebrates outstanding students who display a passion for increasing an awareness of FIRST in their communities. These students are expected to continue to advocate for this program as FIRST alumni. Criteria for the FIRST Dean's List Award include: demonstrated commitment to the ideals of FIRST, a passion for increasing awareness of FIRST in the school and community, an ability to motivate and lead team members, and an overall contribution to the team.

Two students from each team are nominated by their team for this award. From this pool of 6,000 student nominees, more than one hundred finalists for the FIRST Dean's List Award are selected at the competitions. Ten winners of the FIRST Dean's List Award then are selected from the finalists group, with the announcement made at the FIRST Championship.

Brief profiles of the ten 2015 FIRST Robotics Competition Dean's List Award winners illustrate the students' accomplishments as FIRST ambassadors, provide insight into how each of them are changing the culture of their community, and exemplify the power of an individual FIRST student.

Camilo Gonzalez, FRC Team 125, NUTRONS; Boston, Massachusetts

Camilo Gonzales used his participation in FIRST as a springboard to advance the Latino STEM Alliance in Revere, MA, to develop robotics activities in 15 inner-city elementary schools — a task that included teaching computer programming to students who speak Spanish as a native language. Camilo was also a leader on his multi-school FIRST team where he leveraged the strengths of students from three very different schools to create a top performing team.

Alexandra Miller, FRC Team 1629, Garrot Coalition; McHenry, Maryland

Alexandra Miller has been on a FIRST team for eight years, where she initially participated in FIRST® LEGO® League (FLL®) as a fourth grade student. As a FIRST® Robotics Competition (FRC®) team captain she not only directed the team's technical work, but also led the team's effort to engage youth in science and engineering with an innovative project that garnered student interest by using toys as a motivator to learn about internal mechanisms.

Emilie Dufour, FRC Team 3990, Tech for Kids; Montreal, Quebec, Canada

Emilie Dufour is accomplished at getting the message of FIRST out — whether that be to the 30 members of the team she led or the 500,000 viewers who watched her talk about FIRST on a Montreal television show. Emilie took charge of her team's program to provide robotics workshops for physically and mentally challenged children, an effort that was reported in a front page story in the leading Montreal newspaper.

Ben Rowley, FRC Team 1912, Combustion; Slidell, Louisiana

Ben Rowley combined his interests in scouting and robotics by leading his team to support local and state FLL tournaments — an effort that earned him recognition as an Eagle Scout. He also represented FIRST at 30 public awareness events, spreading news about the program to a wide range of people, including politicians, engineers, community leaders, and children.

Sebastian Orellana, FRC Team 4013, Clockwork Mania; Orlando, Florida

As the mechanical lead for his team, Sebastian Orellana helped his team develop award-winning innovations and later applied that expertise to mentor teams in the Junior FIRST® LEGO® League (Jr.FLL®), and the FIRST® Tech Challenge (FTC®). He advocated for FIRST across his school district and beyond by participating in panels and workshops, with his impact reaching as far as Turkey where he participated in a program to expose Turkish students to robotics.

Alyssa Garcia, FRC Team 1684, The Chimeras; Lapeer, Michigan

Alyssa Garcia has seven years of FIRST experience and used her expertise in computer-aided design (CAD), machining, and electrical engineering to teach her teammates. She also used the entrepreneurship and leadership skills acquired while participating in FIRST to establish and mentor FLL teams, and applied these same skills to run FRC Team 1684 as a business with clearly defined roles, responsibilities, and communication channels for each team member and sub-team.

▸ *The FIRST Robotics Competition Dean's List Award recipients were selected based on their leadership abilities, technical skills, and demonstrated commitment to expanding the impact of FIRST within their local community.*

Michael Uttmark, FRC Team 2980, Whidbey Island Wild Cats; Oak Harbor, Washington

Michael Uttmark led efforts to engage his team in using its resources to help others, including manufacturing and repairing items for its school system and for members of its community. He was active in creating a sustainable *FIRST* team with his efforts in this area, including mentoring younger students interested in robotics, leading workshops on soldering, programming, and sensors, and incorporating the engineering design process as a standard operating procedure for decision making.

David Earle, FRC Team 830, The Rat Pack; Ann Arbor, Michigan

David Earle was not only skilled in robotics, but was also an accomplished educator having developed a six-week curriculum to introduce students to CAD, instructed his team on product data management software, and created a workshop to teach elementary-age children about gear ratios. In addition to founding and mentoring an FTC team, David was also a lead designer in a team project to create a robotics laser tag game that taught members about new technologies.

Logan Hickox, FRC Team 4118, Roaring Riptide; Gainesville, Florida

Logan Hickox advanced his robotics team's success by adapting a technique used by his football coach — namely analyzing videos of *FIRST* competitions to identify winning strategies and areas for improvement. As a dual-enrollee at his high school and the University of Florida — where he studied robotics and computer-aided design — Logan developed expertise that benefitted him as the robotics team president and as a mentor to many FLL teams.

Cynthia Erenas, FRC Team 4964, LA Streetbots; Los Angeles, California

Cynthia Erenas exemplifies how students who are provided with the *FIRST* opportunity can thrive as leaders, technologists, and role models. Cynthia helped found her team — located in a challenging neighborhood in Los Angeles, CA, that was not known for technology prowess — and used that experience to develop FTC and FLL teams throughout the community, thereby providing the *FIRST* opportunity to many others.

FIRST Dean's List Award students change the culture, one community at a time.

▶ *The Dean's List Award trophy, presented to each award winner, links the FIRST symbol to its historic origin. Using a balance, Archimedes demonstrated the combined volume of a sphere and cone equals that of a cylinder when the shapes share a common diameter and height.*

CHAPTER 5
COMPUTER-CONTROLLED MACHINING WITH MILLS AND LATHES

The control and automation of manufacturing equipment has increased the productivity and effectiveness of designers. Automation has decreased the entrance barriers to manufacture high tolerance components, thereby providing more individuals with access to advanced manufacturing capabilities. These modern methods of machining include computer-controlled lasers, mills, lathes, and other cutting mechanisms. Though each machine is unique, all require a digital file of the object that will be manufactured, with that file created from a conversational format on the computer numerical control (CNC) machine, or simply entered as a set of computer-aided manufacturing (CAM) instructions. Using a common set of instructions, multiple copies of the same part can be efficiently and exactly replicated, a feature that often benefits robotics applications where multiple copies of the same part are needed. The case studies in this chapter illustrate how CNC mills and lathes have been used to produce components for electromechanical systems. These examples detail how computer-aided design (CAD) models are used to convey design decisions into the manufacturing process and facilitate the subsequent testing and improvement of the manufactured components.

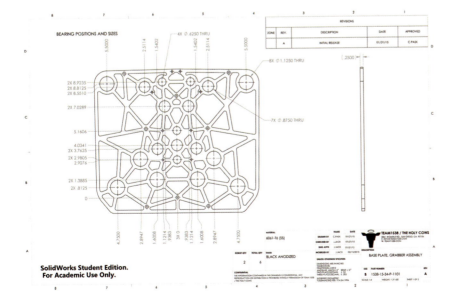

Computer-Controlled Machining with Mills and Lathes | 133

Team 236 – Mastering CNC Mills and Routers

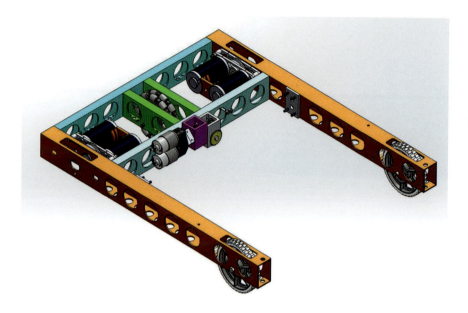

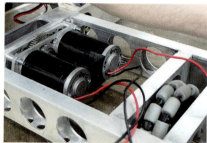

Each of the two front traction wheels was chain-driven by two CIM motors with a 5:1 gear reduction.

A three-wheel drive configuration enabled ground contact with all wheels while the robot drove over the scoring platforms.

TOOLS THAT REDEFINE DESIGN AND MANUFACTURING METHODOLOGY

FIRST® Robotics Competition (FRC®) Team 236, the Techno Ticks, have been building competitive robots since 1999. Since then, this Old Lyme, CT, team has continued to grow and now includes over 10% of the high school student body. The team possesses a passion for science, technology, and all things robotic. The Techno Ticks' 2015 robot was easily recognizable by the large Cat in the Hat decal on the back panel and the swinging "Dr. Seuss Arm," an important feature for its recycling container management strategy. Although the robot's name was "The Tick," the illustrative exterior shared a uniqueness with the beloved children's story, which was, perhaps coincidentally, written using exactly 236 distinct words.

The Techno Ticks' robot was developed with a conscious awareness of available machining capabilities. Two computer numerical control (CNC) mills and a CNC router opened up the design possibilities to create a sophisticated and competitive machine. Evidence of the well-planned design can be seen in almost all robot components. The team established an objective early on to excel in the autonomous period, and every mechanism supported this goal in some way. A clear vision, along with creative thinking, contributed significantly to this team's success as the winner of the New England Waterbury District Event, winner of the Quality Award sponsored by Motorola, and as a finalist at the New England *FIRST* District Championship.

SETTING OBJECTIVES AND REFINING DESIGN

The Techno Ticks began brainstorming immediately after the game was revealed. One of the first actions the team members took was to establish objectives for how the robot would play the game. To be competitive throughout the entire match, the team would build a robot that could consistently build a three-tote stack during the 15-second autonomous period and create multiple stacks of six totes, capped with recycling containers, during the teleoperated period. With a strategy established, team members

○ A set of forks with hinged flaps were mounted on an elevator system and used to stack totes inside the U-shaped chassis.

○ The forks were mounted to a chain-driven sliding carriage that was raised and lowered with two BaneBots motors.

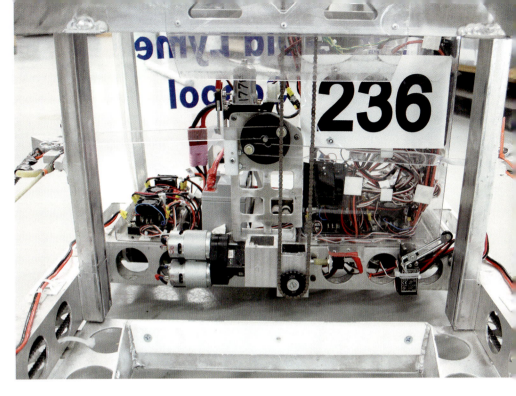

began to concentrate on how to transform these desired actions into detailed physical components. The final robot design incorporated creative ideas contributed by all team members. Once basic drawings of the desired mechanisms were produced, they were refined and sketched in SolidWorks, a solid modeling computer-aided design (CAD) tool. As the design of the individual robot parts evolved, part assemblies were created with the software and then combined to create a model prototype of the full machine.

THREE-WHEEL DRIVE CONFIGURATION AND ELEVATING FORKS

The first systems prototyped were the drivetrain and electronics, mounted to a simple wooden frame for preliminary testing. Shortly after the season's start, the team also constructed wooden mockups of the tote loading station and scoring platform. RECYCLE RUSH℠ was different from previous FRC competitions, as there was no need for defensive capabilities against physical contact from opposing alliance robots. The Techno Ticks realized that the robot wouldn't require an overdesigned robust structure, or need inherent pushing power. They instead opted for a maneuverable drivetrain that could effortlessly navigate the two-inch-tall scoring platforms and mitigate stability issues when transporting tote stacks over and around them. A unique three-wheel drive configuration was selected to maintain ground contact with all wheels through the entire match, thus maximizing control of the robot's position and preventing unanticipated movement. The two front traction wheels were chain-driven by four CIM motors with a 5:1 gear reduction. An unpowered omnidirectional wheel was centered in the back of the chassis, which provided stability and benefitted the robot when making sharp turns.

The mechanism used to stack the totes was a set of forks mounted on an elevator system. Two vertical aluminum tracks supported a sliding carriage between them, to which the fork arms were mounted. The chain-driven carriage was raised and lowered along the two uprights by two BaneBots RS-775 motors with a 4:1 planetary gearbox and a custom 10:1 worm gear reduction to prevent backdrive. A hinged flap mechanism was attached to the inside of each fork to pick up totes. The elevator lowered the forks over the lip of a tote and then reversed direction. The hinge mechanism would catch on the lip of the tote and lift it. This process was repeated until the robot had built a stack of six totes. The forks were also capable of lifting recycling containers, so the stacks were built directly under the container, with the resulting product a capped six-stack of totes ready to be delivered to the scoring platform.

EVOLUTION OF THE ACTIVE TOTE INTAKE

To facilitate collection of totes and recycling containers from the playing field, the team equipped the robot with an active intake system. Several prototypes of this system were built

⬆ A working prototype of the chassis was built to test feasibility of collecting totes from the loading station.

⬆ The initial concept for the active intake utilized BaneBots wheels and motors mounted to the ends of arms that extended forward from the robot.

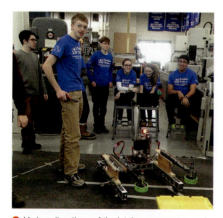

⬆ Various iterations of the intake were prototyped and tested to find the ideal configuration.

for testing various configurations of wheels, belt drives, and arm mounts to find the most efficient way of ingesting the game pieces. The initial concept utilized two BaneBots wheels attached to fixed arms that extended forward from the robot chassis. The wheels were powered by BaneBots RS-550 motors and were separated by approximately 17 inches, or the width of a tote. When the wheels spun in opposite directions, they would pull in or push out whatever game piece they came in contact with. The wheels and motors were mounted at the ends of angled Lexan polycarbonate sheet. To prevent the motors from being damaged, the team designed three-dimensional (3D) printed conical structures to use as protective covers. These cones also served the purpose of guiding totes into the stacking mechanism from the loading station.

The intake system worked well at the team's first district event competition, but there was room for improvement in the area of efficiency. A redesigned intake had arms that opened and closed by a window motor and levers. This design allowed adjustment of the separation between the intake wheels, so the robot could pick up containers and move tote stacks more efficiently. The BaneBots wheels were replaced with Sure-Grip urethane drive rollers. Teardrop-shaped sidewall holes allowed the rollers to be compressed,

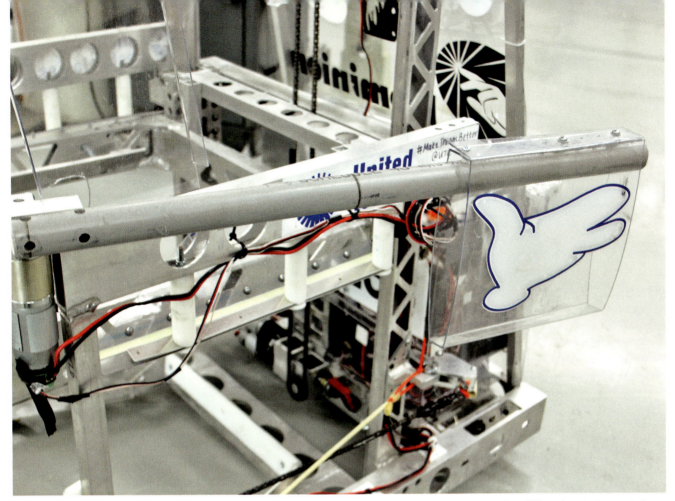

A swinging "Dr. Seuss Arm" was developed to move recycling containers out of the robot's path during autonomous operations. With unobstructed access to the yellow totes, the robot could easily collect them with the active intake system.

thus providing a more secure grip on the game pieces. The RS-550 motors were relocated away from the rollers and were instead connected by a belt drive. This reduced the risk for damage and eliminated the need for the protective cone covers. To replace the tote-guiding functionality of the cones, the team attached KONG® rubber dog toys to rotating aluminum rods above the wheels. With the intake arms closed, the unique shape was perfect for pulling in totes as they were fed through the tote chute.

AUTONOMOUS FUNCTIONALITY

The Techno Ticks' strategy required that the robot reliably build three-tote stacks and move them into the designated zone during the autonomous period to score additional points. However, at the start of the period, the three yellow totes placed on the field were separated by recycling containers. These containers would obstruct the totes, so the team developed a swinging "Dr. Seuss Arm" to swat the containers out of the robot's path, making the totes available to the intake.

The autonomous functionality of FRC Team 236's robot relied on different sensors and software code. The drive gearboxes were connected to encoders to measure the position of the wheels, velocity, and acceleration of the robot. With knowledge of the robot's maximum speed and acceleration rate, coupled with a set of coordinates defining the ideal travel path, the team created a program to calculate the required angular velocity of each wheel and the expected travel distance at 0.01-second intervals. Proportional-integral-derivative (PID) and feed-forward control algorithms were then used to meet the calculated velocity, acceleration, and position goals for each drive wheel.

ADVANTAGES OF USING A TORMACH CNC MILL AND SHOPBOT CNC ROUTER

FRC Team 236 fabricated its robot at the high school's machine shop using assorted advanced manufacturing capabilities, including CNC machining,

A Tormach CNC mill was used to fabricate bearing bores for the drivetrain gearbox plates. This machine is capable of cutting aluminum, steel, and plastics.

Detailed models of the custom gearbox plates, created using SolidWorks software, supported rapid manufacturing on the CNC mill.

The team developed a program to control the speed and heading of the robot during the autonomous period.

milling, and 3D printing. The team could easily manufacture parts on the CNC machines in part because of the detailed SolidWorks robot model generated by the students. The most-used CNC machines were the two Tormach PCNC 1100 personal CNC mills and a ShopBot PRSalpha series CNC router. These tools gave the team the ability to rapidly produce precision parts for the robot.

A Tormach PCNC 1100 three-axis CNC mill can cut aluminum, steel, and plastics. It has a 34-inch by 9.5-inch table and can be operated both manually and automatically. FRC Team 236 used this mill to fabricate the bearing bores for the custom drive gearbox plates, a task that would have been more time consuming and difficult without the CNC capability. These plates made possible the very compact transmission that fit inside the chassis profile, below the electronics. A standoff plate for the elevator transmissions was also fabricated on the CNC mill, designed and machined in the same day. This plate supported an easily removable elevator transmission assembly. The machine was also used to create

Conical protective covers were designed to protect the exposed intake motors.

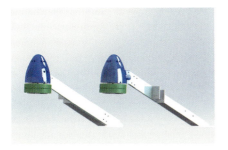
Models of the covers were created to fabricate the final pieces with a 3D printer.

KONG® dog toys were attached to the intake wheels to guide totes into the robot as they exited the chute at the loading station.

A Shopbot CNC router was used to make prototypes of the protective cones for the intake. The plastic prototypes were made from vacuum-formed foam cutouts.

specially shaped pockets on the sliding elevator carriage that held the bearings by the inner races only.

The ShopBot PRSalpha series gantry-based CNC router has a 96-inch by 48-inch table for large-scale machining of various materials, including wood, plastics, and aluminum, and can reach cutting speeds of up to 600 inches per minute. This machine was used to cut weight reduction patterns in the larger structural parts of the robot, such as the chassis and elevator uprights. It was also used to cut foam and wood for prototyping. The 3D-printed cones on the first iteration of the intake were prototyped using foam cut on the ShopBot. The foam cutouts were then vacuum formed to create plastic sheets in the shape of the model for validation before committing to fabricating a 3D-printed version.

Toolpaths for the CNC machines were generated using a SolidWorks software module called HSMWorks, which stands for high speed machining. The Techno Ticks used this software to generate complex toolpaths that were optimized to reduce cycle time without the need to export the CAD models to separate software. Since this software was relatively easy to learn, most team members were able to machine parts as needed. The students were responsible for modeling all of the systems in CAD and for making the CNC parts.

▲ *A three-wheel drive configuration allowed for both maneuverability and stability on the uneven playing surface. The elevator-mounted forks could lift both containers and totes to build capped six-stacks within the robot's frame.*

▲ *The team's number and weight reduction patterns were cut into the elevator support structure using the ShopBot CNC router. The ample work area of this gantry-based machine accommodated the larger robot components.*

▲ *Water was sprayed to lubricate the ShopBot cutting bit, dissipate heat, and remove shavings from material being cut.*

THE IMMEASURABLE VALUE OF CNC CAPABILITY

FRC Team 236 found the biggest advantage to having access to the CNC mills and router was that it allowed students with relatively little machining experience to use these tools with the CAD models to quickly make accurate parts — an important asset with only six weeks to complete a robot. These manufacturing capabilities have influenced not only the team's manufacturing processes, but also have redefined how team members approach design. Without the restrictions of traditional tools, the team can design more creative and effective robots. If a component requires large complex sheet metal parts, the team can fabricate these on the ShopBot router. If intricate shapes, plates, or brackets are necessary, the team can machine these on the Tormach mill.

Unlimited design possibilities result in increasingly sophisticated robots, providing the students with a deeper educational experience. This serves to meet the team's goal: for students to be so inspired by their participation in *FIRST* that they become confident and passionate science and technology leaders. The Techno Ticks' infectious spirit and passion for learning continues to spread, as they show that, truly, "it's in our blood."

▶ *Evidence of the team's advanced manufacturing capabilities could be seen throughout the robot. A clear vision and strategy, combined with CNC machining knowledge and experience, resulted in an award-winning robot.*

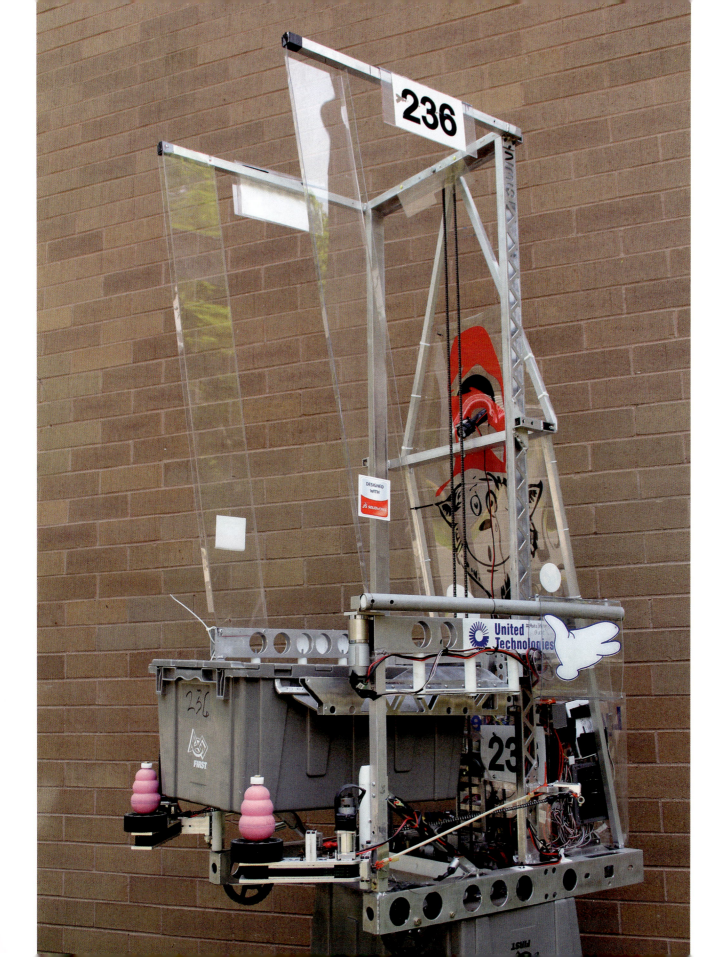

Team 696 – The Circuit Breakers: CNC Extraordinaire

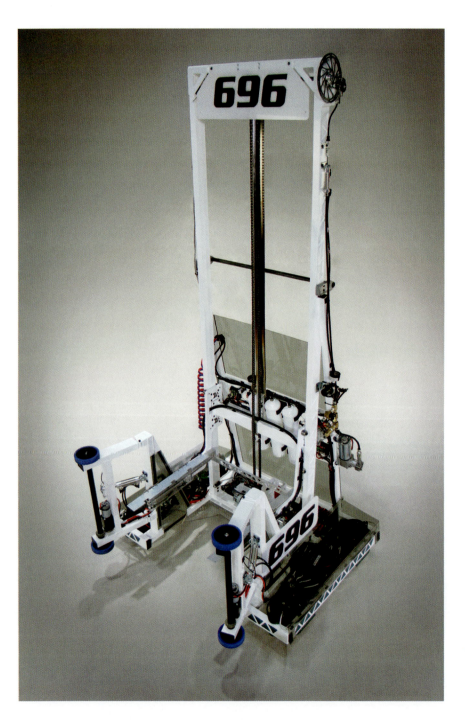

COMPUTER-AIDED DESIGN AS A BACKBONE

The design and manufacturing process used by FIRST® Robotics Competition (FRC®) Team 696, the Circuit Breakers, was one modeled on processes used in industry. Over 80 components on the robot were manufactured using computer numerical control (CNC) machines. A CNC mill was the workhorse to create the drive system and carriage plates for the robot's elevator. A CNC router was programmed to perform pocketing operations on the extruded rectangular tubing for the frame and electronics plates to reduce weight and improve the robot's aesthetics. With all of this work accomplished by students, it was a massive education and execution process.

CNC manufacturing is much more complicated than simply loading a computer-aided design (CAD) file and starting an automated cutting tool. Computer-aided manufacturing (CAM) requires careful design to ensure that the part can be manufactured with an automated process. A complete review of the manufacturing plan must be conducted to be sure that the plan is within the machine's capabilities and operating envelope. Proficiency in a number of software packages to design parts and program the CNC machines is needed to be successful in this domain. FRC Team

CNC manufacturing was critical to produce the elegant robot named "Centurion." Each element of the robot's structure and mechanisms was manufactured using CNC technology, with some mechanisms containing dozens of unique parts.

696 developed a unified process that tracked each component from its idea to fabrication, with a full review of the part at a few points during the process to be certain that the design achieved its needed purpose.

MANY STEPS FROM CAD TO CAM

Autodesk Inventor software was used to create both conceptual and detailed designs. The finalized robot CAD model was the product of more than 100 revisions and featured hundreds of individual components and numerous assemblies. To help manage this process, Autodesk Vault was used to coordinate the collective workflow, control the revision process, and ensure the integrity of individual parts and integrated systems. The Vault File Server preview browser aided the process as it allowed team members to step through the revision history for parts and assemblies and view their evolution during the entire design cycle.

The design process started with prototypes being created for key systems such as the elevator and intake mechanism. For the drive system, published designs from previous competitions were studied to establish initial concepts. Once the critical dimensions of these primary systems were determined, the information was passed to the CAD sub-team to design the needed mechanisms. Concepts and initial ideas were refined using the Microsoft program Zoomit to mark up CAD models and highlight areas needing revision. As an example of one of the benefits of the preliminary review process, it was determined that a 50-tooth gear on the drive system would make contact with the scoring ramp in specific orientations. To address this situation a 44-tooth gear was used in place of the larger gear, a change easily made during this part of the design cycle.

○ The swerve drive assembly was one of many systems designed by the students and fabricated in the team's workshop using CNC mills and lathes, as well as other manufacturing tools and machines.

Another example illustrated the value of a detailed CAD process. Weight was accurately estimated once the material density properties were defined in the CAD model for each component. Pocketing and other lightening strategies were used for components having excess material, with the process reducing the components' weight. The CAD model was also useful to place electrical components virtually and plan for the integration of these elements with the rest of the systems on the robot.

At the conclusion of this iterative process, the digital designs were finalized and reviewed by a group of team mentors. Once approved, the CAD files for each mechanism were passed to the CAM and animation sub-teams. The animation sub-team imported the files into Autodesk 3ds Max software to apply material styles and animate the systems to create a high quality exploded view of each assembly. This process was critical to present the complicated and compact assemblies in a format that could be used to assemble the manufactured pieces into a working system. To reduce the rendering time that could total 30 hours of computer processing to produce one minute of animated video, Autodesk Backburner was used to create a distributed rendering farm using 50 otherwise idle computers. Using this virtual rendering farm of networked computers, the single-computer 30-hour process was reduced to one hour.

While some parts could be fabricated by hand from two-dimensional (2D) CAD drawings derived from the solid models, the majority of parts were manufactured using CNC equipment. The conversion of the CAD files to instructions that could be processed by the CNC machines was completed by the CAM sub-team. The primary responsibility of the CAM sub-team was to be sure that every part could be manufactured in the team's manufacturing lab. The CAM and CAD groups worked closely throughout this process. To be sure the designed parts had the correct dimensions and could function as intended, the lead mentor reviewed all parts before the CAD files were transferred to the CAM sub-team.

Students take charge of every process, including CAD modeling, programming CAM toolpaths, and operating CNC machines. The Haas Mini Mill and its advanced control system transformed the team's design and manufacturing processes.

CAM software specific to each CNC machine was used to convert the CAD files into a series of toolpaths and cutting instructions for each part. These toolpaths were visualized within the CAM software in the form of three-dimensional (3D) simulations to be sure machine limitations were adhered to and that the created part matched the original solid model. The CAM toolpaths, tool feeds, and cutting speeds were carefully reviewed by a pair of students, followed by an independent review by two mentors, before the instruction set advanced into production.

To track this process, a fabrication sheet for each machining setup was created to record key parameters such as the part number, a list of tools and offsets, location of the part's work coordinate origin, and the operational speeds, feeds, and maximum vertical depth. Once this information was recorded, the instructions were produced as numerical control G-code that could be interpreted by each CNC machine. As an example of the complexity of these instructions, the team's largest numerical control file contained nearly 150,000 lines of code and totaled 4MB in size.

The final step of the in-house manufacturing process was executing the G-code in a CNC machine. For FRC Team 696 those machines included a Techno-Isel 30-inch by 30-inch CNC router and a Haas Mini Mill vertical machining center, with the latter serving as the primary workhorse for producing the team's precision-machined components.

Using a virtual training system, students built proficiency in Haas CNC mill programming and operations, with skilled operators earning an industry-recognized certification.

Once the numerical control instructions were loaded into the Haas CNC mill, a graphics simulation was used on the mill's control interface to verify that the program did not violate any of the machine's operating limits and that the designed toolpaths did not exceed the machine's range of motion. Once the raw stock was positioned in the mill and its location verified, the first manufacturing process was run with the machine set to 5% of its maximum motion speed. This slow speed allowed time for the student operator to stop the manufacturing process if the operation was not correct. After

○ The robot's frame was pocketed using a CNC router. A cooling mist and dust collection system lubricated the router bit and removed aluminum chips. In addition to saving weight, the triangular pattern enhanced the robot's appearance.

○ The weight reduction pockets produced a triangular pattern on the frame members. The original tube's wall thickness was 0.125 inches and the pockets were machined to a depth of 0.095 inches. The pockets retained a 0.030-inch-thick bottom and provided an illusion that the pieces were machined from solid bar stock.

the first part was produced and its dimensions verified, subsequent copies of the same component were manufactured, with the machine running at 100% of its maximum speed.

The ability to manufacture parts in-house using CNC equipment completely transformed the team's design process. Almost every part for the robot was designed from the onset with the expectation that it be fabricated with a CNC router or mill. Accomplishing the CNC process on an enclosed machining center such as the Haas CNC mill not only allowed for precision and accuracy, but also provided for high spindle speeds, coolant use, and automatic tool change — each of which maintained quality and benefited a rapid production schedule.

Once parts were manufactured they were sent off site for powder coating and anodizing at a sponsor's facility. Aside from this step in the process, students conducted the majority of the design and manufacturing work in the team's high school shop, with mentor interactions provided when needed. The design and manufacturing process of the robot's base and elevator carriage plates illustrates the values of the CAD to CAM to CNC machining process.

DESIGNING AND FABRICATING

The robot's base and elevator carriage plates were manufactured using a CNC mill and CNC router, with the latter used for pocketing operations on extruded aluminum tube frame members. Other structural components, such as a lightweight plate to mount electronics, were also manufactured using automated cutting equipment.

The base was assembled from 0.25-inch-thick aluminum extruded tube that was lightened with pockets cut into the external frame. The machined rectangular tubes were then welded together to construct the robot frame. The router was also used to machine the robot's electronics panel as a lattice structure to reduce the platform's weight. Holes were drilled into this plate and later tapped to directly mount electronics equipment on the plate, thereby avoiding the use of additional fastening hardware.

The CNC mill was used to machine the two elevator carriage plates at the base of the elevator. The automated milling process began with the creation of the CAM toolpaths, with different colors illustrating the material cuts (green), high speed cuts (blue), and rapid transverse motions to new cutting locations (orange). Animations of the cutting sequence provided a visual means to verify the operation before material was actually cut.

Once the toolpaths were verified, the machining was initiated — in this case cutting both plates from the same piece of 0.5-inch-thick plate to reduce waste and minimize machining time.

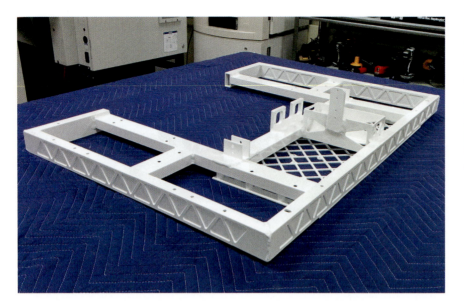
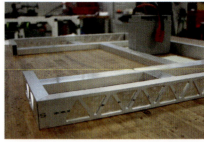

The frame rails were arranged as a temporary mockup to test fit the battery prior to welding. The TIG-welded frame included all tabs and platforms. Welding avoided the need for threaded fasteners and provided a sturdy foundation for the robot's mechanisms.

The raw material was mounted to a fixture plate that was secured to the mill to allow for the plates to be fully cut out. Milling the original 16-inch by 12-inch plate to produce the two carriage plates took approximately one hour on the Haas CNC mill. Once the plates were cut and powder coated, they were installed on the robot frame as a sturdy foundation for the elevator.

Two mechanical systems on the elevator reduced the load on the elevator motors. A large constant force spring was installed at the top of the elevator to offset the weight of the elevator carriage. With the additional force provided by the spring, the mechanism accelerated upward at a higher rate. To help preserve the motors, a commercial brake rotor was mounted in line with the elevator drive shaft. When needed, pneumatically operated brake calipers gripped the rotor and held the load. This braking system prevented the lifting motors from operating in a stalled condition to maintain the elevator's vertical position.

SWERVE DRIVE MAGIC

After extensive analysis of the game, the team decided that a swerve drive maximized the robot's agility and increased its overall performance. The swerve drive would allow the robot to perform complex maneuvers such as translating while simultaneously rotating. This system required each wheel to be independently powered and steered using separate motors and transmissions for each action. The team designed this drive system with automated machining in mind. Each drive required a number of components to be machined, including the housing for the transmissions as well as the upper and lower plates of the drive unit.

The first step in the process was to create a CAD model of the swerve drive to establish the packaging for the two rotational systems within a confined space. Given the uniqueness of this configuration, a three-dimensional (3D) full-scale model of the design was printed and assembled to ensure all parts fit as expected into the compact space. A MakerBot Replicator 2 3D printer was used to construct this physical model using polylactic acid (PLA), a biodegradable polymer.

The primary structure that housed the bearing blocks, gears, axles, and bearings was named the "swerve tube." The part was manufactured from a 3.5-inch-diameter solid cylinder of aluminum. Over 90% of the original piece of aluminum was machined away, reducing the starting mass of one kilogram to a manufactured mass of 99 grams. Manufacturing this component required machining from six sides and was a complex operation, with the CAM programming alone taking a week to complete. One hour was required to machine each swerve tube. As a single element, the swerve tube included a number of functions that otherwise would have required additional parts. Despite being a complex design and milling operation, the swerve tube saved time since it eliminated the need to design, manufacture, and assemble many other individual parts.

Other components of the swerve drive were also manufactured using the Haas CNC mill. Each fabrication process began with visualizing the cutting routine and was completed by post-processing the manufactured parts for installation. For the upper and lower plates, post-processing included anodizing the plates. Off-the-shelf components were also machined for the drive system, including 84-tooth gears used for steering. For these parts, which served as the structural foundation of the drive mechanism, a groove was cut along the perimeter to install 60 ball bearings. Also, a clearance pocket was machined for the wheel and its drive gear. As with other components, the CAD model was animated to create a realistic view of the mechanism to guide the assembly and installation process.

A PLACE TO CREATE

FRC Team 696 is based out of a 3,000-square-foot design and manufacturing lab at the Clark Magnet High School in La Crescenta, CA, a space created over the last four years.

The space was designed with the objective of building student proficiency in manufacturing processes, with an expectation that students experienced in manufacturing would reach higher levels of success in engineering design. Prior to 2012, the team worked out of a small storage room and did not have any significant equipment or machinery.

The lab currently houses two Haas CNC vertical machining centers, three CNC lathes, two CNC routers, and seven 3D printers. The lab also includes traditional machining equipment, assembly areas, and a computer lab. Adjoining areas accommodate a site for electronics and programming work, additional equipment space, a welding station, a CNC plasma cutter, and storage. In addition to nearly 50 students involved in robotics, 100 other students in the school's three engineering and manufacturing courses also use the lab.

The lab facilitates collaboration because all of the sub-teams are collocated. With the students responsible for the CAD, CAM, and

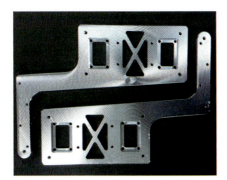

The swerve tube was the primary structure in the drive system and housed the drive's bearing blocks, bearings, gears, and axles. Six separate machining operations were needed to manufacture each of the four tubes used in the drive system.

The toolpaths that produced two elevator carriage plates from a rectangular aluminum plate illustrate how material cutting motions were used to fabricate parts.

◉ *Machined parts were placed in a vibratory tumbler with ceramic material to remove burrs, rough edges, and machining marks. Green anodizing added to the vibrant appearance of the swerve drive upper plates.*

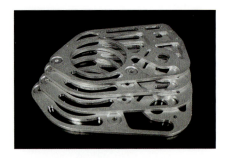

◉ *The swerve drive integrated all aspects of the drive system, including the propulsion and steering motors and their transmissions, into a single component.*

machining of all parts, communications and training were an important aspect of the team's work. At the beginning of the season the students were guided by mentors on the use of the CAM software and CNC machining processes, but as the season progressed the students were able to program and manufacture parts with minimal oversight. Part of this success with student proficiency can be attributed to the school's CAM curriculum, which leads to industry-level certification in CNC programming and operations. Also, a rigorous fall training program for the team allowed students to become proficient in a chosen skill area by December, and minimized the need for students to learn new skills during the build season.

MORE THAN DESIGN

Expert use of CNC machining produced an elegant and aesthetic design that resulted from a rapid development cycle. Attention to detail ensured high levels of precision and quality in the manufactured parts. The level of customization using in-house CNC manufacturing processes far surpassed options available using off-the-shelf components or manually machined components.

Machining components using CNC processes necessitate that a team invests heavily in the time needed to become proficient in CAD modeling and CAM programming prior to manufacturing. Although parts designed with the intent of CNC manufacturing require a larger initial investment of time in the design and setup stages, the investment pays dividends in the automated manufacturing process, and especially so when multiple quantities of the same part are needed. Also, the completed parts match the design plan nearly exactly, allowing for ease of assembly and integration into the robot. When a well-designed plan is manufactured with a high level of quality, the overall performance can almost be guaranteed.

◗ The swerve tube, tube cover plate, and bearing block components of the drive system were manufactured by students using a Haas Mini Mill CNC vertical machining center.

Team 1538 – The Holy Cows: Leading the Herd in Computer-Controlled Machining

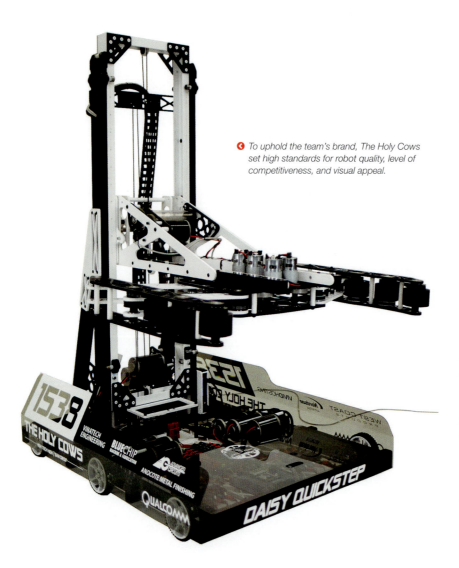

To uphold the team's brand, The Holy Cows set high standards for robot quality, level of competitiveness, and visual appeal.

DIVERSE SPONSORS PROVIDE UNIQUE MACHINING CAPABILITIES

Branding is an important part of the strategy of any business or corporation — it is a means of self-promotion and distinguishes one from competitors. A consistent brand communicates a certain level of professionalism and can help set expectations. This holds true not only for traditional businesses, but also for FIRST® Robotics Competition (FRC®) teams. A solid marketing strategy and positive image projection are among the many things in which FRC Team 1538, The Holy Cows, from San Diego, CA, excels.

The Holy Cows have built a reputation as true professionals when it comes to promoting the team's brand. This easily recognizable team has established and published standards not only for graphics and apparel, but also for its robots. Along with specific templates for team number, robot name, and sponsors, the robot must also adhere to strict appearance standards, "to make our competition robot visually appealing." These include a requirement for material finishes that all aluminum parts should be powder coated gloss black or gloss white, or anodized in black, Type II or Type III, depending on the purpose of the part being anodized.

To successfully meet such high standards, the team must rely on advanced machining to generate the detailed aluminum components and on facilities to apply the coatings. Founded in 2005, FRC Team 1538

🔴 Much of the robot manufacturing was performed by the students, who were trained in shop safety, CNC programming, and use of various machines at a local sponsor's machine shop.

has spent the past 11 years expanding its computer numerical control (CNC) manufacturing capabilities, which include access to four CNC mills, a CNC lathe, and a waterjet through support of its multiple sponsors. These resources have been invaluable in maintaining a reputation for high standards and an image of consistent quality the team has worked to create.

CLOSED LOOP DESIGN EVOLUTION

To maintain a competitive edge, a successful FRC robot design must sometimes adapt as the competition season progresses and game play evolves. While one response is reactionary, truly effective teams instead actively pursue continuous improvement of their robot and its capabilities. FRC Team 1538 has adopted the philosophy of a proactive approach and uses it as a foundation for what the team calls a closed loop design process. This methodology is composed of five basic stages that continuously repeat: strategic analysis, prototyping, computer-aided design (CAD), manufacturing, and assembly/test. Once the process begins at the season's start, it doesn't end until the team's last match at the *FIRST* Championship. The team members exist in a constant state of self-analysis and continuous improvement of the robot throughout the season to maintain a competitive advantage over other teams. The interminable evolution and improvement of their robot led team members to claim that they "never run the same exact robot configuration twice during a season."

The most significant design changes occurred during prototyping while the team was deciding on a strategy. To support the desire to have an innovative robot that was unique to any previous robot designs, The Holy Cows tested many prototypes to unveil which solutions worked and which didn't. Once the general robot concept was formed, follow on changes were less drastic, yet still beneficial. As evidence to FRC Team 1538 never competing with the same robot twice, between the team's two regional competitions and the *FIRST* Championship, it made changes to add functionality, increase speed, and improve scoring performance.

Initial game simulations led The Holy Cows to consider an internal tote stacking robot, however upon analyzing further, the team arrived at another solution. FRC Team 1538 believed it could improve on the idea of a robot that built six-stacks of totes from the loading station or landfill by designing a robot that could quickly stack and deliver two totes at a time to the scoring platform.

MULTIPLE SOURCES OF ACCESSIBLE TOOLING AND MACHINING CAPABILITY

The majority of robot manufacturing happened at team sponsor BlueChip Machine and Fabrication, Inc., where the team spends nine to 14 days out of the build season working on the robot. Here the students are empowered to do the machining themselves, working under mentor supervision. Each year, up to 10 students from different grades are sent to learn about shop safety, CNC programming, machining, cleaning parts, performing quality assurance checks, and preparing components for welding, powder coating, and anodizing. After gaining experience, most students are able to operate the machines independently and begin to train other team members. A mix of ages and experience levels maintains a constant supply of trained students and ensures that at all times there is at least one student who knows how to operate each machine. Mentor involvement is limited to assigning jobs, coordinating workflow, and assisting with machine programming. One of the mentor employees at BlueChip performed all welding on the 2015 robot.

At BlueChip, the team has access to three two-axis CNC mills, which are used for machining aluminum tubing and for all post-waterjet operations, including counterboring, hole tapping, and opening up holes. Traditionally, the team cut aluminum tubing with a hand saw. However, as the robots became increasingly intricate, team members found it easier to use a CNC mill to machine the desired angles into the tubing.

BlueChip also has a two-axis CNC lathe and three manual lathes with digital readouts, primarily used by FRC Team 1538 for manufacturing all shafts, spacers, and standoffs. The Hardinge and Feeler manual lathes have little runout and were used for precision work. A Webb manual lathe was used for shape changing processes such as broaching, parting, and facing on pulleys and gears.

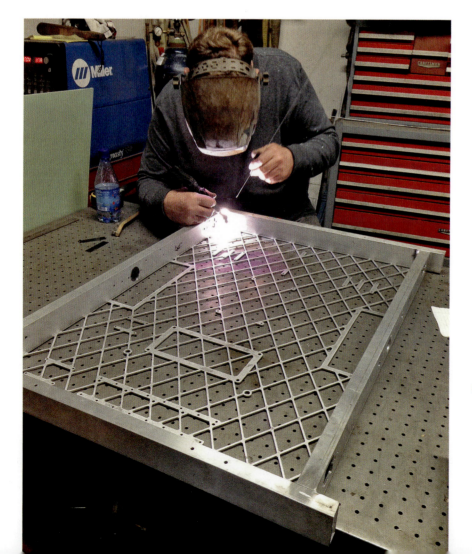

◀ *Students prepared parts for welding, which was performed by a mentor. Through its multiple sponsors, the team had access to four CNC mills, a CNC lathe, a waterjet, and a laser cutter. The use of advanced manufacturing techniques was essential to create the detailed aluminum components.*

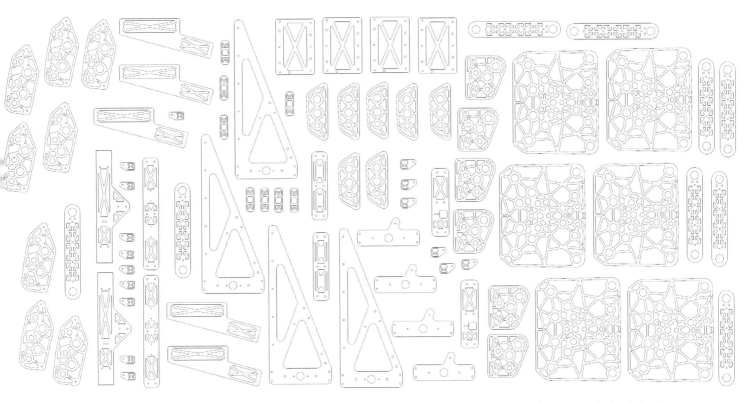

▶ A Flow Hyperjet waterjet was used to cut robot components from large pieces of sheet metal. The components were nested to use as much of each sheet as possible during a single cut. Bearing bores and counterbores were machined on a mill to achieve the required tolerances.

Vinatech Engineering, Inc. is another sponsor of FRC Team 1538, lending the machining capability of a Flow Hyperjet waterjet and a laser cutter. The waterjet was used to machine materials of varying thickness, from 0.0625 to 0.125 inches thick. The team used a combination of material acquired from a local vendor's donated remnants and from purchasing the material in four-foot by eight-foot sheets. Vinatech acquired the laser cutter shortly after the 2014 FRC season, and the team's parts for the 2015 robot were some of the first to be made on the new equipment. Although the laser cutter could machine to tight tolerances, The Holy Cows experienced limitations with this machine due to the intense heat buildup, which caused warping of thinner materials on components such as the electronics baseplate. The team learned to only use the laser cutter for parts with 0.25-inch thickness. Overall the laser cutter reduced manufacturing time by eliminating rework and the need to construct fixtures to hold the pieces, as team members had to do with a mill.

At the school shop, The Holy Cows had access to a two-axis CNC mill and a manual lathe. The mill was used during prototyping and the final manufacturing phase for reworking or modifying parts. For complicated pieces that the team was unable to machine themselves, including bearing blocks used on the drivetrain and elevator system, team members were able to send these parts to Qualcomm Inc. for fabrication at its machine shop. All powder coating was applied at RW Little, Co. and all anodizing was done at Anocote Metal Finishing, Inc. These local companies typically had a two- to four-day turnaround for a completed product.

The capabilities provided by advanced machining equipment enabled the team to design and build increasingly sophisticated robots in less time. The first robot built by FRC Team 1538 only incorporated a few custom pieces, but the 2015 robot incorporated over 140 unique parts. The team recognizes that the ability to manufacture customized components for the robot directly supports development of solutions specific to the challenge.

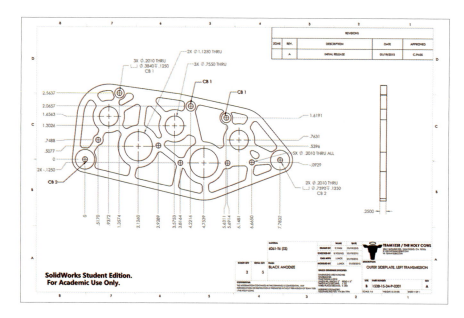

The robot was modeled using SolidWorks software, from which two-dimensional drawings were created. The drawings called out specific dimensions, tolerances, the material, and the material finish. Each component was assigned a unique part number for tracking and assembly.

COMPUTER-AIDED DESIGN TO GENERATE PART DRAWINGS

In order to fully utilize CNC capability, the design process must focus heavily on CAD. The Holy Cows used SolidWorks software to model the entire robot and SolidWorks Workgroup PDM to control design revisions and facilitate collaboration. For the components that required CNC machining, the team used the model to generate drawings for the specific parts. Three different drawing types were generated: part drawings were used to machine stock material, waterjet files were used to turn sheet metal into cut shapes, and weldment drawings were used to weld components. It was not uncommon for a part to have more than one drawing generated for its construction. The gearbox side plates required both part drawings and waterjet drawings. The plates were cut using the waterjet to rapidly create the pocketing patterns, but then had to be placed on a mill for machining bearing bores, counterbores, and tapping holes. These features required tight tolerances unachievable on the waterjet.

THE ROBOT, TOOLING, AND CONTINUOUS IMPROVEMENT

The Holy Cows' sleek powder coated robot relied on two key components to score points: a claw to grab totes and recycling containers and an elevator to lift these objects. The initial game strategy was for the robot to build stacks on the scoring platform in three stages: two totes, then another two totes, and then one recycling container. After its first regional competition, the team identified a way to improve scoring performance. While practicing, team members found that the robot could reliably lift one tote with a recycling container on top if the claw incorporated antennae for stabilization. The revised strategy made a slightly taller stack, thus scoring more points. The revised three-stage strategy became two totes, plus two totes, plus one tote with a container on top. The team continued to improve the effectiveness of this system through the *FIRST* Championship.

The unique claw was a single mechanism designed to quickly acquire and manipulate both totes and recycling containers from the landfill or loading station in any orientation: totes and containers right side up, on their sides, upside down, and different tote widths depending on which side was approached. The claw used a powered intake to essentially pinch the totes and containers. Prototypes of the arms in the parallel linkage of the claw were made on the CNC mill at the school to test different configurations for strength. Each side of the final claw was made from two parallel aluminum plates, machined using the waterjet. Sandwiched between the plates were three timing belt-driven rollers to draw the totes and containers into the claw.

The claw was used to grasp both totes and recycling containers. It was opened and closed using a four-bar linkage and had powered rollers to assist with game piece acquisition.

The rollers were powered by two RS775 motors that, for protection, were located away from the ends of the claw. A single gear and series of 20 timing belt pulleys arranged at the pivot points on the four-bar linkage produced a motor speed reduction of 80:1. The pulleys, manufactured on a lathe, and timing belts were hidden inside the four-bar linkage to conserve space. A powered pinching function that produced over 205-foot-pounds of torque firmly gripped the totes and containers. The two sides of the claw were quickly opened and closed by a mechanism consisting of a four-bar linkage, machined on the waterjet, powered by two dedicated RS775 motors with a 320:1 reduction gearbox. The gears in this gearbox were pocketed by students using the CNC mill, removing 50 to 60 percent of the original weight.

A two-stage, cable-driven elevator system was implemented to quickly raise and lower a carriage – the structural component the claw was mounted on. The first stage of the lift drove an inner white frame to extend upwards along the outer black uprights, sliding in bearing blocks mounted to the frame. The second stage drove the carriage to extend from the bottom of the white frame to the top. The system extended to 71 inches in less than two seconds, and could lift up to three totes or tote/container combinations at a time. To keep the elevator frame pieces square and prevent warping, they were held together with gusset plates and riveted rather than welded.

The team used pulleys and AmSteel-Blue synthetic rope for the cable system that raised and lowered the elevator components. The AmSteel-Blue had a working load of 1,400 pounds and was a fraction of the weight of steel cable. A dual-motor and gearbox assembly wound the cable around a lightweight spool made from a two-inch diameter, 0.125-inch wall thickness length of shatter-resistant polycarbonate tubing with acetyl hubs pressed into the ends. A spring-loaded cable slowed down the inertia of the spool and prevented

A CNC mill was used to pocket the gears for the claw, reducing the weight by 50 to 60 percent.

Computer-Controlled Machining with Mills and Lathes | 155

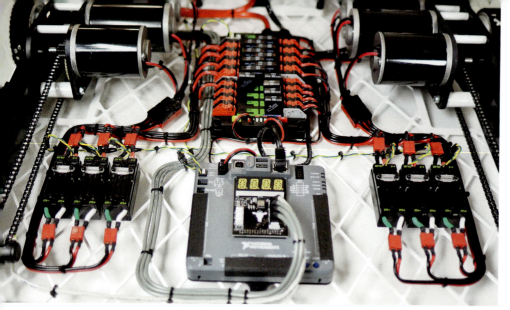

● Each side of the six-wheel drivetrain was powered by two CIM motors and a MiniCIM. Methodical placement of the electrical components and wiring on the chassis provided easy access for maintenance and modifications. The uncluttered design was part of the robot's overall visual appeal.

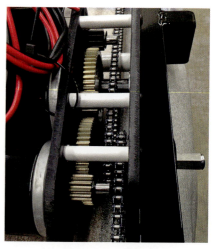

● The custom drivetrain gearbox side plates were powder coated black and the standoffs were powder coated white.

backlash of the cable when the elevator reached the end of travel.

The selected strategy of stacking only two totes at a time required exceptional robot maneuverability. With a top speed of 18.5 feet per second, FRC Team 1538 definitely achieved this, as the robot's name, Daisy Quickstep, indicates. The six-wheel drivetrain, powered by four CIM motors and two MiniCIMs, drove four-inch powder coated AndyMark high strength aluminum performance wheels. Black treads made of conveyor belt material were secured to the wheels with white rivets. A special battery box was designed with a built-in passive lock to prevent the battery from being inadvertently disconnected during matches. The layout of the aluminum chassis, welded at BlueChip, was well planned and had clean, accessible electronics and wiring. The uncluttered, elegant layout of the final machine contributed to the iconic look of this team's robot.

THERE'S ALWAYS TIME TO IMPROVE

One of the few maneuvers in RECYCLE RUSH℠ that could affect the opposing alliance's score was to take possession of the recycling containers on the step. The alliance with the extra containers had the ability to score higher, and prevented those points from going to the opponent. After its first regional competition, FRC Team 1538 saw the benefit of this capability and added fast dual arms capable of reaching out and grabbing two recycling containers from the step during the autonomous period. The use of creative machining to reduce component weight created a margin of 15 to 20 pounds, which enabled the addition of this feature. Between the team's second regional competition and the FIRST Championship, it enhanced these arms to make them even faster.

Access to the CNC machines and waterjet technology greatly influenced how The Holy Cows' team members designed their robot. Part design must take into account manufacturing resources — better resources result in more complicated parts, built faster and with higher quality. The ability to design weight reduction features into a component's structure was crucial to creating margin for making robot modifications and improvements throughout the season. This highly functional robot excelled at the 2015 challenge, leading the team to become finalists in two regional competitions. The robot's unique look also didn't go unnoticed, earning the team multiple awards for industrial design, quality, and imagery. FRC Team 1538 members' manufacturing experience helps them consistently build high quality, competitive, and aesthetically refined robots, often exceeding the high standards they set for themselves.

● Daisy Quickstep featured over 140 custom fabricated parts. CNC machining enabled fast manufacturing of high quality, complex assemblies that were part of the robot's sophisticated and iconic look.

Team 3250 – Efficiencies with Automated Manufacturing

EVERYBODY CADS, EVERYBODY CAMS

One benefit that every student member has on FIRST® Robotics Competition (FRC®) Team 3250 is participating in their school's four-year Manufacturing and Design curriculum. The program includes instruction in computer-aided design (CAD) and computer-aided manufacturing (CAM), augmented with hands-on training using state-of-the-art equipment. This experience provides each student on the team with valuable CAD and CAM skills. In addition, each student can operate the automated machinery to convert digital models into physical components. This depth of design and machining talent has produced a team where students have manufactured all of the robot's parts, with mentors providing oversight of the design process. Each part was designed not only to be functional, but also to enable the part to be efficiently manufactured in the team's workshop.

SolidWorks software was used to design the robot and its individual components. HSMWorks software converted the digital parts into instructions that were interpreted by the automated machinery to select the correct tools and sculpt raw material into the designed part. The suite of computer numerical control (CNC) equipment available to the team at the high school work site included a Haas TM-1P vertical mill. This fully enclosed machine has an automatic ten-station

All components of the gearbox for the robot's arm were designed in CAD and manufactured using a Haas Vertical Toolroom Mill. CNC milling produced the gears, sprockets, and bearing plates for this multiple stage transmission.

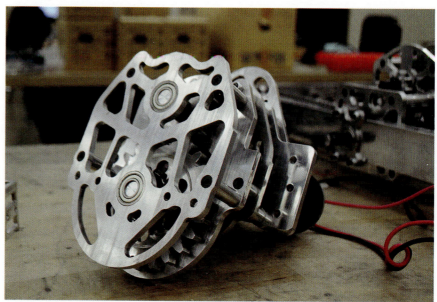

Each gearbox was assembled and installed on the robot as a composite component. With this design, the entire arm lift system, including the motor, transmissions, and power take-off sprockets, could be replaced as a single unit.

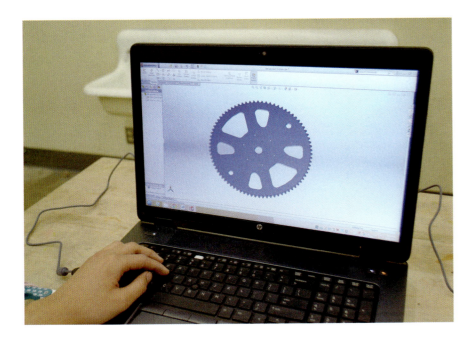

A sprocket created as a CAD sketch was converted into a solid model before being manufactured on a CNC mill. The manufactured part was deburred and finally installed as an assembly on the robot.

tool changer and a working volume of 3.3 cubic feet, a size that influenced the team's design decisions. Other equipment included a manual mill and lathe, a drill press, a plasma cutter, grinders and sanders, and PROBOTIX FireBall Comet CNC routers for cutting plastic and wood. The combination of student talent, mentor expertise, advanced software, and automated as well as manual manufacturing tools provided ideal conditions to design and fabricate an FRC robot.

ROBOTICS: PUTTING THE CURRICULUM INTO PRACTICE

Manufacturing considerations were a key aspect during the team's design process. Armed with concepts, team members drew whiteboard and paper sketches to share and advance their ideas. Once preliminary ideas were refined in the hand sketches, CAD models were created and a formal design review was conducted to solidify the elements in each subsystem. The digital models of the subsystems were assembled into a single CAD model as a proof of concept for the composite system. The exactness of the CAD model facilitated a careful review of all components to ensure the interoperability of the subsystems and eliminate interferences between components.

Once the components were verified in the composite CAD model, the design of each part was examined from a production perspective to minimize the manufacturing time. This included reducing the initialization time and machining toolpaths, as well as finding the best methods to secure the raw stock in the automated mill. During this review, parts were altered to speed up the machining process and decrease the chance of manufacturing defective parts. For example, the forward claw to grip containers and totes was modified to optimize machine fixture requirements. The size of the backplane on this gripper was also altered to fit this part on the bed of the mill. Other parts had features added for the sake of easing manufacturing

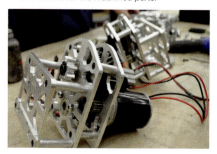

The drivetrain included custom-designed CNC gears that provided the robot with the needed gear reduction to achieve a specified speed. The CAD model doubled as assembly instructions for the machined parts.

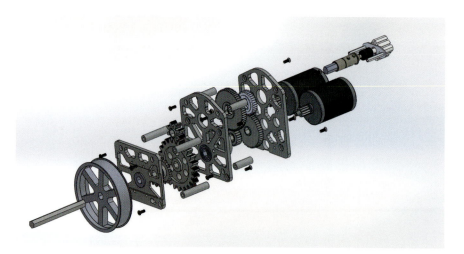

such as a design change to the drive wheels. Flat contours were added on the inside of each wheel to create an attachment point for a set of custom jaws that held each wheel during machining operations.

Without addressing these issues during the detailed design phase, the dormant problems would linger and require that a solution be found when the parts were manufactured. Eventually the hidden issue would surface and slow down production. By putting great effort into the design of parts with an awareness of the team's manufacturing capabilities, FRC Team 3250 was able to eliminate potential manufacturing problems at the earliest stages and avoid design-related production delays.

To be sure that each part could be efficiently manufactured by the team, each design was reviewed and approved by the team's head of design — a student who was very skilled at using each of the team's automated manufacturing machines. Only after this approval process did the part progress to the next stage where the students responsible for that subsystem developed the CAM instructions to cut the parts out of raw material. The CAM toolpaths were verified using video simulations to virtually review the manufacturing process before any material was machined.

Some parts benefitted from additional CNC-assisted prototyping reviews to increase the team's understanding of complex systems. Developing exact prototypes of proposed systems using wood instead of aluminum allowed configuration and geometric details to be more quickly perfected. In addition to saving time, using wood was also cheaper than creating aluminum prototypes. The forward claw was prototyped in this manner to increase quality and eliminate errors in the final design. A CNC router was used to cut plastic templates of each of the claw's structural members. The templates were used to trace the patterns on plywood sheets that were then cut using a jigsaw. This process highlighted the team's ability to combine advanced and basic manufacturing processes to quickly construct high performing working systems.

MOVING FROM THE COMPUTER TO PRODUCTION

With all of the resources available at the high school's workshop, including design software and extensive machining hardware, the process to progress from computer models to machined parts that were ready to be assembled was fast and efficient. The availability of CNC tooling saved the team significant time while manufacturing components and especially when making multiple copies of the same part. For example, the drive system required two identical transmissions to be manufactured with some of the same parts used multiple times in each transmission.

For such systems, the time saved using CNC manufacturing methods was even more significant. When multiple versions of the same piece had to be manufactured, the team-initiated production runs with the mill were devoted to making multiple copies of a single part at one time. This process minimized set up time and capitalized on the machine operator's ability to develop proficiency making multiple copies of the same part.

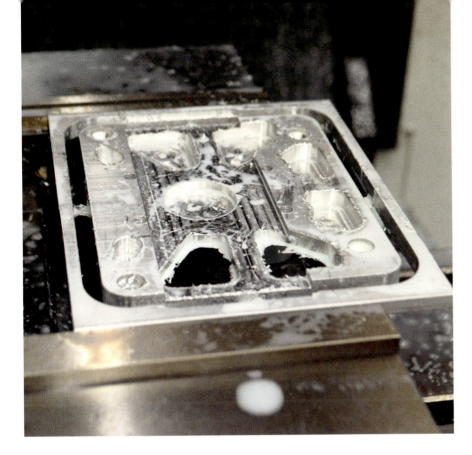

Caution and direct oversight were used during the initial run of each part. During each first run, the student operator maintained a careful watch on the cutting processes for unproven toolpaths.

Similar to the production run methodology to manufacture parts, an industrial process guided the assembly process. Once the parts were approved for manufacturing, the group of students that designed the parts manufactured all of the parts for its subsystem. The manufactured parts were cleaned, inspected, and inventoried. When all of the parts were available, an assembly line was created to attach the components together and create working subsystems. The subsystems were then inspected and tested before being combined with other components to create the robot.

Aluminum and plastic were the materials used almost exclusively to construct the mechanical and structural components of the robot, with 93 percent of the parts made with aluminum and the remaining seven percent made with plastic.

The aluminum components included the drive base, a front claw, a six-bar arm to support the claw, and a rear gripping mechanism. The team also manufactured all of the sprockets, transmission gears, and bearing blocks. The plastic components manufactured by the team included chain-tensioning cams, mounts for the pneumatic system, encoder gears, and a core material for a carbon fiber guard on the gripping mechanism. The Haas mill was used to machine 90 percent of the aluminum parts, with the other aluminum parts not machined. All of the plastic parts were machined using CNC routers.

FRC Team 3250 manufactured at least twice as many parts as needed for one robot, using the second set of parts to create an identical robot. The second robot became the practice and test platform to refine driver skills and explore new design ideas after the build season's end. With all of the parts manufactured on CNC machines, the two robots were identical twins and had the same performance characteristics. The exactness of the two robots greatly improved the team's level of preparedness between the last day of the construction period and the first day of actual competition.

FROM HAND TOOLS TO AUTOMATION

The team's proficiency using automated machine tools greatly advanced in 2015 due to the acquisition of a Haas TM-1P vertical mill. While the team previously used hobby-grade machinery to fabricate wood and plastic parts, prior to 2015, FRC Team 3250 did not have advanced CNC equipment. Without a CNC mill capable of cutting metal, the team was restricted to cutting aluminum with a hacksaw and drilling

All of the manufactured components were inspected before being assembled into functional systems. Extra pieces were fabricated during the manufacturing processes and designated as spare parts.

CNC machining eased the ability to manufacture multiple components and build twin robots: one serving as an evaluation and programming platform and the other being the refined robot used in competitions.

aluminum with a drill press. The vertical mill was a major investment for the Manufacturing and Design program at the high school, but the investment quickly manifested itself as a meteoric advancement in student manufacturing skills. The ability to manufacture a robust, high quality robot for the 2015 FRC season marked a quantum leap in the team's capabilities and the students' manufacturing proficiencies.

By incorporating machining considerations into the initial aspects of the design process, FRC Team 3250 was able to quickly and efficiently create high quality parts in a short period of time. The ability to design and manufacture those parts, including multiple parts within the team's workshop, produced a very reliable and high performing robot. As one example of quality, the team's robot did not encounter any mechanical failures during the three competitions in which the team participated. That is quite an achievement for a team that spent the prior year working with just a drill press and hand tools.

◆ CNC milling was used to manufacture many components on the robot, including structural members and motor transmissions. The team's mastery of CAD and CAM allowed custom gears and sprockets to be designed, fabricated, and used on the robot.

Team 4293 – Custom Components for Competitive Robots

◠ *A Fadal vertical mill was used for milling aluminum tube stock. The team worked closely with professional machinists, who mentored the students through demonstrations and hands-on learning.*

◑ *The team's workspace is attached to an aerospace-grade CNC machine shop. One of the tools available to students is a Mori Seiki MV-55/50 mill, which was used to perform the majority of the machining.*

EXPERIMENTING WITH CNC CAPABILITIES

The accessibility of modern computer numerical control (CNC) machining tools has inspired *FIRST*® Robotics Competition (FRC®) Team 4293, Team Komodo, from Highlands Ranch, CO. This team has capitalized on the ability to transform a computer model to a high tolerance component, evidenced by the many custom machined parts featured on its robot. While the team acknowledges the use of these manufacturing tools as an advantage to the robot's competitiveness, it also sees the students applying their creativity and developing critical design skills. Team Komodo has developed a close partnership with professional machinists and strives to constantly reinvent how it can apply CNC machining to robotic design.

WORKING CLOSELY WITH PROFESSIONALS

FRC Team 4293 has the good fortune to work closely with two local aerospace-grade CNC machine shops — so close that the team's workspace is actually attached to one of the shops. This unique relationship between students and professional machinists and equipment has helped the team define itself and its robots through creative application of CNC capabilities.

Expert machinists at both shops are mentors to the students, helping them develop machining and strategic thinking skills. They teach the students through demonstration and encourage hands-on learning, allowing the students to assume the role of robot fabricator. The students on Team Komodo are empowered to lead the manufacturing efforts, from developing the parts using computer-aided design (CAD) tools and uploading the CAD drawings to the CNC machines, to performing the

◐ Factors to consider when designing components for CNC machining include the physical properties of the material, functionality of the final product, and time to perform the machining.

◐ The robot was modeled using PTC Creo to evaluate individual component integration.

machining steps and producing the final parts. This high level of student engagement is augmented by the expert machinists, who are available to provide guidance and mentoring.

The RECYCLE RUSH^{SM} robot parts were produced on a variety of machines made available to the team by the shops, including four Mori Seiki vertical mills and one Fadal vertical mill, a Sharp vertical knee mill, a Bridgeport manual mill, a Mori Seiki lathe, and Okuma manual lathes. The majority of machining was performed on a Mori Seiki MV-55/50, an older three-axis vertical mill, known to be forgiving to the student machinists-in-training. The team also had access to a Brown and Sharpe coordinate measuring machine (CMM), which was used to recreate spare parts if detailed drawings weren't available.

FROM CARDBOARD TO CAD

The team's design process began with analysis of the scoring system and game components. The intent of this exercise was to identify a strategy that would be most effective and beneficial to potential alliance partners. The team members agreed that to be effective in this game, they would build a robot that's primary function was to collect totes from the landfill. The details of the design soon became clear and a forklift-inspired configuration emerged. The team then employed a preliminary form of prototyping. Cardboard components were constructed and tested, and the parts that worked were transitioned to full-scale paper cutouts for validation. Team Komodo used this method to narrow down design options instead of the CNC capability, which was reserved for final fabrication only. It reduced the CAD time required to model the many proposed designs, reduced the number of parts to machine, and reduced the time required for equipment calibration.

Once the final robot began to emerge, individual components were modeled

◐ *A five-hook system was used to lift game pieces. Each hook was attached to a slider bar that rode up and down vertical tracks. The final iteration was powered with a single CIM motor and gearbox.*

using PTC Creo. Exact dimensions and specifications of components were established, and then the individual parts were joined to create a single model of the final robot. The merging of the components helped student designers to visualize potential issues with weight or size limitations, as well as range of motion restrictions, prior to spending any time in the machine shop. Validation of the final design and dimensional tests ensured that the final product represented the desired solution. The students worked closely with the machinists on the final CAD design. These experts taught the students to identify design changes that could increase the durability or machinability of a part, while maintaining its original intended functionality.

Once the CAD models had been finalized, they were imported to Mastercam, a CAD tool with an integrated set of predefined toolpaths to guide CNC machines. Mastercam then exported a G-code (a numerical control programming language) toolpath to the particular machine that was needed. The students calibrated the machines, with oversight from a professional machinist, and performed test runs before machining each component.

FRC Team 4293 identified five key factors for each member to consider when designing CNC components: the physical properties and machinability of the material, the location of the piece on the robot, the functionality of the final product, the machine size and capability, and the time to perform the machining.

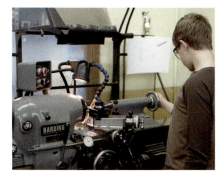

Custom spacers that separated the gearboxes and frame were manufactured on the manual lathe. The chassis and frame were fabricated from aluminum tube stock machined on the Fadal vertical mill. Students were immersed in all manufacturing processes.

A FULLY CUSTOMIZED LIFTING MECHANISM

The robot's identifying feature was a forklift-inspired, tote lifting mechanism, which used an innovative five-hook system that could either lift five totes from their top lip individually, or three totes capped with a recycling container. The individual control of each lifted component eliminated the design concern of all weight resting on the lowest tote in the stack during transport. This design also enabled the robot to build onto smaller stacks already placed on the scoring platform.

The team selected a hook design after experimenting with several off-the-shelf hooks from a local hardware store. As the design evolved, the team used a three-dimensional (3D) printer to prototype hook variations to accelerate testing without using valuable machining time. The final hooks were attached to five horizontal slider bars that straddled two parallel vertical rails. Each slider bar was fitted with hooks on the outer edge and low friction nylon blocks infused with molybdenum attached to the ends of the bars to facilitate sliding up and down the rails. The vertical rails were fabricated from a solid square aluminum extrusion, which was machined on the Fadal vertical mill into an I-beam shape. The decision to manufacture custom rails was made when the team couldn't find a stock item that met their design criteria.

The nylon blocks were selected to reduce friction between the slider bars and vertical rails. They were precision milled on the Mori Seiki MV-55/50 using two end mill sizes to match the I-beam shape of the vertical rails. Testing revealed that under the weight load of a tote, the nylon blocks would seize on the rails with any slight misalignment. To address the issue, the team had to redesign the portion of the slider bars in contact with the vertical rails, while minimizing changes to the other lift components.

The solution came in the form of roller bearings in place of the friction-based sliding approach. The roller bearings, designed to work with the existing vertical rails, were created from milled steel blocks with stud type track rollers. The new design was drawn in CAD and then quickly produced on a CNC mill for testing. The steel

◆ Roller bearings were added to the slider bars to improve the vertical rail interface. The redesign was modeled using PTC Creo before being fabricated on a CNC mill.

◆ The heavier steel roller bearings were an improvement over the sliding nylon blocks in the original design.

bearings were significantly heavier than the nylon blocks, but the team found this to be a benefit because gravity helped lower the slider bars when releasing totes on the scoring platform instead of needing the motor to pull the friction-based sliders down.

Initially, two CIM motors and a gearbox powered the tote lifting function. Early in the competition season the team found that under heavy loads the motors were unable to maintain their position and the tote stacks would slowly slip downwards. A last minute design decision was made to use a planetary gearbox with a 1:64 gear ratio instead. This enabled removal of one CIM motor, negated the risk of back drive, and eliminated the danger of burning out the motor controllers.

The chassis and frame were milled from two-inch by one-inch aluminum tube stock with a 0.125-inch wall thickness, selected for its weight and durability. This particular material also supported the use of rivets for fast assembly of structural pieces and mounting of components. The chassis components were milled to precise lengths and assembled exactly as anticipated from the CAD model.

EXPLORING CNC POSSIBILITIES

CNC offers precision manufacturing for parts and components that integrate seamlessly. With the ability to machine to exact dimensions, components assemble effortlessly and spares can be easily made with accurate reproduction.

FRC Team 4293 is especially proud of its robot's unique CNC-machined elements. CNC introduces the possibility of limitless components, which drives the students to think about robot design differently. Each year the team members learn more about machining capabilities and continue to increase the complexity of manufactured parts to support their unique designs. The team members' excitement to learn and to push the limits of integrating computer-controlled machining features on their robots ensures they will be presenting increasingly innovative solutions to the ever-changing FRC challenge.

◆ Each tote was lifted and lowered separately. The vertical motion of the slider bars was improved with redesign of a critical component. The modified part was quickly modeled and fabricated using CNC machining, then implemented on the robot.

FIRST® ROBOTICS COMPETITION MAKERSPACES

Collaborations to Design and Create

The maker movement, a do-it-yourself trend that has become more prevalent in the past ten years, is revolutionizing traditional learning methods. The availability of open-source technology and affordable manufacturing processes has expanded the maker culture to a new group of innovators — high school students. A makerspace is a community-based facility where people share resources and knowledge, and collaborate to make things. FIRST® Robotics Competition (FRC®) teams utilize the same structure when designing, building, and testing their robots. Three FRC makerspaces stand out as exemplary facilities that could become models for future teams to emulate.

In 2005, Jackie Moore, a now-retired Systems Engineer from Chicago, IL, was interested in enhancing her children's education to increase their exposure to science and math. She wanted to provide them with a more technology-based, hands-on learning environment, and found that FRC addressed many of the areas about which she was concerned. At the time, there were few Chicago-area teams, so in 2006, Jackie took the initiative to start five FRC teams, including FRC Team 1739, the Chicago Knights.

After years of working in temporary spaces, the team finally found a permanent home in the Ford City Mall, previously a factory for several car companies. Jackie named the space Level UP, which was initially located in the mall's basement, but was soon relocated to the main floor. The shared workspace is home to the Chicago Knights, as well as FIRST® Tech Challenge (FTC®) teams, FIRST® LEGO® League (FLL®) teams, and other STEM-based after school programs. Adult mentors are available to provide advice and support if needed, but the learning environment is mostly peer-based. The public space attracts students from a multitude of backgrounds and locations around Chicago, and the doors are open to anyone who wants to take part in the hands-on learning activities. The facility is ideal for outreach activities and spreading the message of FIRST, as well as exposing more students to STEM.

At just under 350 teams, Michigan has more FRC teams than any other state in the United States. FRC makerspaces can provide a large number of students with access to tools and shared resources, as well as collaborative learning environments, in one location. In 2010, in an effort to increase community involvement and develop Detroit's future leaders, the University of Michigan College of Engineering collaborated with Detroit Public Schools to create the Michigan Engineering Zone (MEZ), located in the University of Michigan's Detroit Center. The MEZ was established to inspire Detroit's youth to pursue an education and eventual career in science and technology. The 5,200-square-foot facility, provided by the University of Michigan College of Engineering, houses computer labs, a machine shop, a practice playing field, and collaborative workstations. Engineers from local industry and University of Michigan faculty, students, and alumni mentor the students. The space is also used to educate students and their parents about college opportunities and applications, financial aid, and available scholarships. During the 2015 FRC season, a total of 17 Detroit high schools, comprising 275 total students, called the MEZ home.

The first year-round FRC makerspace in the United States to be integrated in a college campus is the *FIRST* Robotics Community Center, located at Kettering University in Flint, MI. Opened in September 2014, the previously vacant 9,600-square-foot gymnasium houses a programming lab, a practice playing field, collaborative work areas, individual storage spaces for eight teams, and has an adjacent machine shop. The space was created as a collaborative space for students to learn from each other and from the Kettering faculty, staff, and students. Veteran and rookie teams from diverse backgrounds have a place to work together while pursuing the hands-on learning that is such an important part of *FIRST*. Future renovations are planned to add room for more FRC teams, a conference area, and space dedicated to students involved in FLL. Five teams used the facility for the 2015 FRC season, including the Kettering-sponsored team, FRC Team 1506, Metal Muscle.

As participation and interest in the maker movement grows, more and more people are able to share their knowledge, experience, and devotion to lifelong learning. It should be no surprise that *FIRST* is a part of this exciting new trend. The growing popularity of *FIRST* makerspaces takes advantage of the inherent community-based learning and collaboration that all teams exhibit. *FIRST* team members gain exposure not only to their own mentors, but to members of other teams with whom they are co-located. These co-located teams are not only designing, building, and testing robots. As Jackie Moore describes it, they are also "contributing to and learning from the entire local *FIRST* family of teams."

FIRST makerspaces are facilities where groups of teams meet to collaborate, build robots, and learn from each other. Members of different teams share makerspace resources, including shop equipment, tools, and mentor expertise.

CHAPTER 6
COMPUTER-CONTROLLED CUTTING SYSTEMS

This chapter presents examples of using computer-controlled cutting systems to fabricate components for robotic systems. The profiled systems operate in two dimensions and include laser, plasma, and waterjet cutting systems. Similar to the entrance point for computer numerical control (CNC) mills and lathes, the starting point for CNC cutting systems is a digital file that is commonly created using computer-aided design (CAD) software. Since the process of incorporating a digital model is common to all CNC manufacturing methods, a designer or fabricator who is skilled in one manufacturing technique can leverage that awareness to develop the specific skills required for a different machine. The case studies detail how two-dimensional (2D) shapes fabricated on laser, plasma, and waterjet cutters are manipulated, joined, and combined to form complex three-dimensional (3D) structures. The ease of use and speed of CNC cutting systems makes them useful not only for final design fabrication but also for prototyping ideas. Because these systems are relatively uncomplicated to setup and operate, they are ideal platforms for introducing the concept of CNC machining to new designers and fabricators.

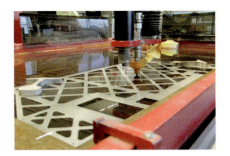

Computer-Controlled Cutting Systems | 173

Team 67 – Heroes of Tomorrow

The 2015 HOTBot was designed with CNC cutting in mind. Waterjet-cut sheet metal components were incorporated in every subsystem of the robot.

CNC CUTTING INTEGRAL TO TEAM'S SUCCESS

Computer numerical control (CNC) cutting can greatly influence robot design, allowing for more flexible and innovative solutions. This manufacturing capability, and particularly waterjet cutting, can create unique sheet metal patterns that would not be possible with traditional machining methods. FIRST® Robotics Competition (FRC®) Team 67, the HOT Team, from Milford, MI, has significant experience with advanced manufacturing. The 2015 "HOTBot" featured many components that were transformed from simple, flat sheet metal parts into complex, lightweight assemblies using a waterjet cutter.

DESIGN PHILOSOPHY INSPIRES WORLD-CLASS ROBOTS

For FRC Team 67, waterjet cutting was used as the main manufacturing method for the majority of its robot's machined parts. Beginning with the first brainstorming session and continuing throughout the design phase, as many parts and assemblies as possible were designed for manufacturing from sheet metal using waterjet cutting. This limited the need for traditional machining, providing better fabrication accuracy and faster turnaround time.

The capability afforded by waterjet cutting influenced the team's design process and became central to its design philosophy. By fully incorporating this fabrication method into its process, the team integrates traditional machined parts with computer-controlled machined parts to produce world-class robots.

The robot was designed to manipulate totes and recycling containers to score points in both the autonomous and teleoperated periods.

A detailed model of the robot was created using SolidWorks software. Holes for mounting and weight reduction were included early in the design process.

Design of the 2015 HOTBot began with a thorough analysis of the game, objectives, and scoring. This helped the team narrow down a list of priorities for the robot, which led to formulation of a strategy and initial design concepts. Throughout this process, the team maintained an awareness of available manufacturing resources to generate a feasible and realistic solution. The final robot was capable of scoring tote sets and collecting recycling containers from the step in the autonomous period, and then used totes from the loading station or landfill zone during the teleoperated period to create at least two capped six-stacks per match.

THE PROVING GROUND FOR WATERJET CUTTING

The HOT Team built its robot in the central machine shop at the General Motors Milford Proving Grounds in Milford, MI. There the team had full access to the tools and equipment, which included CNC mills, manual lathes, sheet metal brakes, and a waterjet cutter. All parts of the robot were manufactured by students with guidance from mentors.

The team had full access to a Flow waterjet in the shop, which was used to cut the majority of the two-dimensional (2D) pieces used on the robot. These parts included frame panels, motor mounting plates, and structural components for the robot's subsystems. FlowMaster, the waterjet cutting software, was used to control the complex machine operations. Different applications in the FlowMaster software suite were used to lay out the parts to maximize use of material, generate toolpaths, and control the integrated operation of

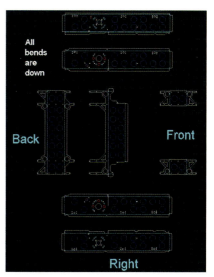

A two-dimensional drawing of the flat panel frame sections was uploaded to the waterjet. Cutting operations were controlled by FlowMaster software.

the pump, cutting head, and cutting table to cut the material. Select students were trained to operate the waterjet under mentor supervision. They were responsible for loading the stock material, uploading the drawing files, arranging the toolpaths, and observing the process to ensure the parts were correctly cut.

MULTI-DIMENSIONAL MODELING

FRC Team 67 approached its detailed design process differently than many teams, using both 2D AutoCAD and three-dimensional (3D) SolidWorks computer-aided design (CAD) software tools for the majority of the robot parts. 2D components were created in AutoCAD, from which students created solid 3D models of the parts and virtually joined them using the 2D assembly sketches as guides.

For generating the manufacturing drawings, the team used a top-down design process. The 2D subsystem assemblies initially created in AutoCAD were used to generate the individual part drawings reflecting the unbent geometry of the part. This was necessary because the waterjet required 2D drawing exchange format (DXF) files of the flattened parts to make the cuts in the sheet metal.

INCORPORATING WATERJET CUTTING IN EVERY SUBSYSTEM

A significant amount of detailed design was performed on the robot chassis and frame. Placement of the holes for gearbox mounts and axles, slots for mating parts, and lightening holes were all incorporated as a part of the baseline structure, which was cut with the waterjet. Aluminum panels for the chassis were designed with flanges, which were bent with the metal brake. When assembled, these bent structures provided the stiffness needed for the open-front design. The accuracy and precision of waterjet cutting allowed the team to integrate tight tolerance features that wouldn't normally be feasible with traditional machining techniques. Although the planning required significant attention early on, the benefit was realized with assembly, which was relatively effortless.

Totes were collected from the loading station with a powered intake that used wheels to pull totes into the robot as they exited the tote chute. The wheels, axles, and bearings were sandwiched between two 0.0625-inch-thick aluminum sheet metal plates, which were connected to the robot chassis via flexible Lexan

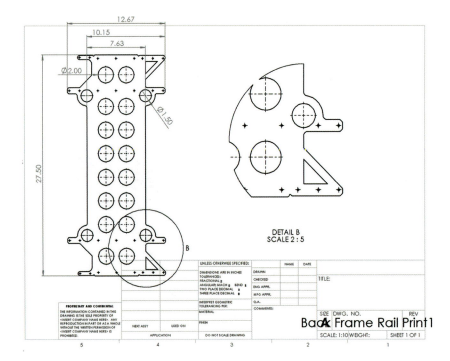

◆ Detailed two-dimensional drawings included accurate dimensions to ensure parts were cut to proper specifications.

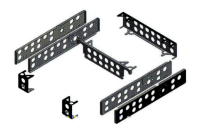

◆ An exploded view of the lower chassis frame assembly illustrates the various sheet metal parts manufactured using the waterjet. The incorporated flanges provided stiffness to the assembled pieces.

A powered intake pulled totes into the robot from the playing field floor and as they exited the chute. Rapid manufacturing and implementation of revisions to the intake assembly were enabled by the waterjet.

polycarbonate sheet. The assembly was designed around the waterjet's capability. All sheet metal plates were created as 2D drawings and then uploaded to the waterjet as DXF files. The mechanism underwent multiple iterations to improve performance, including changed mounting positions, component orientations, system geometry, and wheel materials. The waterjet enabled fast manufacturing of each modification.

A cable-driven internal stacker lifted the collected totes. Once lifted, another tote would be loaded through the chute, and the process repeated until a six-stack of totes was complete. The stacker arms were made from waterjet-cut aluminum plates with Delrin rollers and traveled up and down a tower made from 1.5-inch thin walled aluminum tubes. Additional waterjet plates were mounted to a one-way spring hinge that deflected as the stacker was lowered over the top of a tote. When the stacker was raised, the plate hooked under the lip of the tote. The aluminum stacker plates were designed with flanges so that after being cut on the waterjet, the flanges could be bent. When all of the plates were assembled, the box structure was capable of repeatedly lifting 50 pounds of totes without breaking.

A double-jointed arm was mounted to the back of the robot to collect recycling containers from the field and cap tote stacks. The structure was made from one-inch by two-inch rectangular aluminum tube, cut with a CNC mill. The arm rotated 120 degrees to lift the containers, relying on four large custom 100-tooth sprockets. The sprocket pattern was created with AutoCAD Mechanical's sprocket generator and then sent to the waterjet to be cut.

◔ The recycling container arm components were cut using the waterjet.

◔ The waterjet was used to fabricate the components of the four-bar linkage assembly that grabbed recycling containers off the step.

◔ Students were trained to operate the waterjet, located at the General Motors Milford Proving Grounds. They were responsible for loading stock material, uploading drawing files, arranging toolpaths, and monitoring the cutting process.

Between competitions, the HOT Team added a set of arms to the back of the robot to grab recycling containers off the step during the autonomous period. Each arm was actuated through a four-bar linkage. The frame, mounting plates, linkages, and hooks on the system were all cut with the waterjet. This allowed for the parts to be quickly machined and assembled.

IN HOT PURSUIT OF EXCELLENCE

FRC Team 67 leveraged its machine shop access to build a robot that featured waterjet-cut components on every subsystem. The team's successful integration of computer-controlled cutting tools in the design and build of the 2015 HOTBot earned the robot a Quality Award sponsored by Motorola and an Excellence in Engineering Award sponsored by Delphi at the team's district events. The sophisticated robot also performed exceptionally on the field, leading the team to win two *FIRST* in Michigan district events and become a finalist in the *FIRST* Championship Carson Subdivision.

The students continue to learn new techniques for machining on the existing equipment, and as the machine shop acquires new tools, the team embraces the added capability as part of its process. The Heroes of Tomorrow have been producing exceptional robots and practicing continuous improvement since 1997, and there's no sign of that changing any time soon.

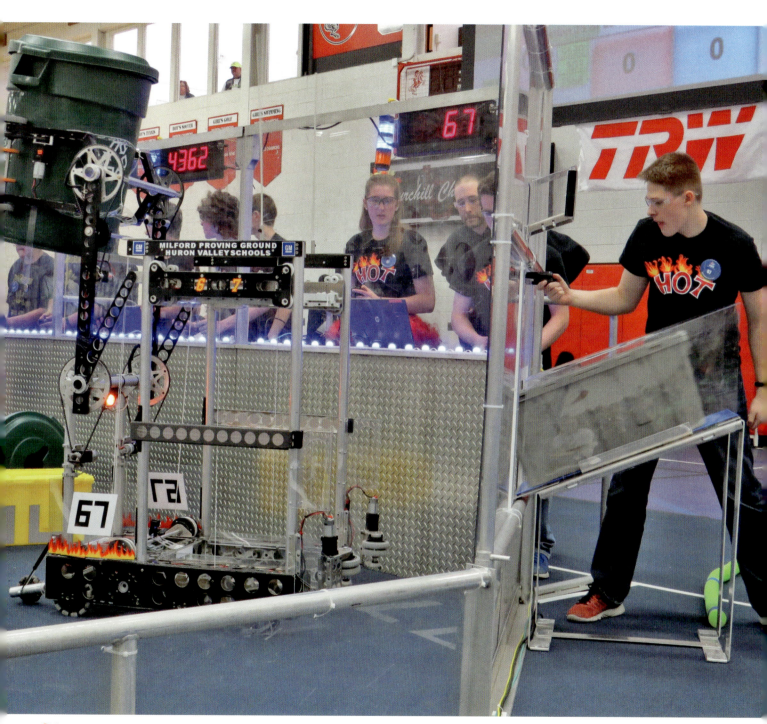

Computer-controlled cutting techniques have greatly influenced the team's design process. The capability afforded by waterjet cutting transformed flat sheet metal into complex assemblies to create a world-class robot.

Team 148 – The Robowranglers: On the Cutting Edge of CNC Capability

◆ FRC Team 148 used its experience with CNC cutting methods to rapidly prototype and manufacture an award-winning robot.

180 | FIRST Robots: Behind the Design | Vince Wilczynski and Stephanie Slezycki

⬢ Upon identifying a high-level goal to excel in RECYCLE RUSH,℠ the team developed a strategy for playing the game and the necessary robot features to achieve success. The resulting machine earned the team many awards for exceptional performance, including winning the FIRST Championship Curie Subdivision.

BATMAN AND ROBIN TO THE RESCUE

The 2015 FIRST® Robotics Competition (FRC®) game introduced a difficult challenge. All qualification, quarterfinal, and semi-final match winners were determined using match point averages, and only the final matches earned teams a win, loss, or tie designation. Using this structure, for teams to get to the finals the robots had to do more than win a match; they had to maximize their alliance score, every time. Reliability became an important robot attribute that could be achieved through intelligent design and quality manufacturing.

One team that understood how to design excellence into a robot using sheet metal and computer numerical control (CNC) cutting capabilities was FRC Team 148, the Robowranglers. This Dallas, TX, team recognized the benefit of CNC, which enables rapid prototyping, manufacturing, and assembly, and supports complex and customized construction. The team's experience and knowledge was essential for success in RECYCLE RUSH℠. Although the challenge was difficult, the Robowranglers team members were relentless in their quest to build an exceptional robot. They set their sights high with a determination to win "every match," a goal not easily achieved. However, with their design and manufacturing skills and constant iteration, the team members achieved two high-performance robots that met their goal.

IDENTIFYING THE HIGH-LEVEL GOAL

The rigorous design process established by FRC Team 148 began with a brainstorming technique known as goal mapping, which is used to identify a high-level goal and the actions required to achieve it. For the 2015 season, the Robowranglers identified that goal to be winning the FIRST Championship. Drilling down a level, the team concluded that to achieve this they must maximize the number of points the robot could score. The next level of detail determined what the robot would have to do to maximize points scored, which was by scoring as many recycling containers as possible on top of the highest tote stacks possible. The team found that this could be accomplished by building either the fastest center step recycling container-grabbing robot, or the best scoring robot. The Robowranglers decided it was wiser to build the best scoring robot, anticipating that the best scorer would be the top seeded robot in competitions and could then select the best container-grabbing robot as an alliance partner for final matches.

Once team members had established their strategy, they began working on ideas for how to achieve it. Imagination

Computer-Controlled Cutting Systems | 181

◉ A prototype of the tote intake was fabricated from plywood to experiment with geometries.

◉ A dual-robot concept, nicknamed Batman and Robin, worked together to build and score stacks of totes and recycling containers.

◉ As the intake was refined, it was mounted to a reconfigurable chassis to determine if it could pull a full stack of totes.

and crazy ideas were encouraged and each concept was given a name, including one called Batman, who "has no superpowers, just does things better than everyone else." The team asked, "How would we make Batman the best scoring robot?" to which someone answered, "We'll give him a sidekick. Every superhero deserves a good sidekick." This was the beginning of the Batman and Robin concept — dual robots working together to outscore the competition.

Robin was a stationary robot with the sole responsibility of building six-stacks of totes in front of the loading station. Tethered to Robin was Batman, a mobile robot that picked up recycling containers, capped the stacks built by Robin, and delivered them to the scoring platform. These two robots utilized parallel path manufacturing to accomplish simultaneous tasks and maximize scoring efficiency.

FROM PROGRESSIVE PROTOTYPING TO DETAILED DESIGN

Initial prototypes were built with cardboard, plywood, and two by four boards to experiment with anticipated challenges, such as tote and recycling container manipulation. As the concepts evolved, refined prototypes and a reconfigurable chassis were constructed to observe system interaction with the totes and recycling containers and to test drivetrain configurations. Preliminary prototyping was also done using SolidWorks software with crude three-dimensional (3D) computer-aided design (CAD) models created to visualize basic component integration, test ranges of motion, and lay out key dimensions.

Robot component machining was provided by sponsor Innovation First Manufacturing, Inc. The CNC equipment used included a laser cutter, turret punch presses, a router, and metal brakes.

Once the team had prototyped initial concepts and collected lessons learned, the detailed design process commenced. Dimensions were measured from the prototypes and transferred into the CAD models, as well as detailed features of other robot components and assemblies. Sheet metal templates were selected that aligned with guidelines set by the team's metal shop sponsor, and then parts were created using the proper templates for material type and thickness. The CAD models included thorough detail of each robot feature, which proved to be useful for rapid assembly, easy integration of new components, and replacement of parts, if needed.

STANDARD PRACTICES FOR MANUFACTURING

FRC Team 148 is sponsored by Innovation First Manufacturing, Inc. (IFMI). The CNC equipment most used to fabricate the team's robots include a laser cutter, turret punch presses, a router, and metal brakes. IFMI provided the machined components on the condition that FRC Team 148 adhere to the shop's standard practices for part design. Examples of these standards include bend parameters, drawing practices, fixed turret punch hole sizes, flange sizes, and proximity of holes to bend lines. The CNC machining resources provided by IFMI shaped the team's design process, which has evolved to take advantage of the available tools and technologies.

The Robowranglers chose to use sheet metal for the majority of the robot structural components because of its fabrication speed, ease of machining, and versatility to make both large assemblies and small complex pieces. Batman and Robin exhibited three distinct styles of sheet metal construction that were made possible with the team's CNC machining resources. Parallel plate construction consisted of flat plates cut to shape then held parallel to each other with standoffs. False extrusion shapes were made from two nested C-channels riveted together to form a square tube. Lightweight shell structures were created to span large sections while exhibiting great strength relative to the thickness of the material.

A LASER CUTTER AND TURRET PUNCH PRESSES MAKE QUICK WORK OF SHEET METAL CUTTING

The robot parts were cut at IFMI with an Amada punch/laser combination machine that couples a laser cutter with a high speed, servo-electric turret punch press. The laser cutter was used to quickly cut sheet metal patterns drawn by the team and had the ability to incorporate accurate mounting and attachment holes in the parts, as well as weight reduction patterns. The turret punch press was used to cut holes in the sheet metal using a standard punch tool set that made three different sized holes. While the turret punch press wasn't as versatile as the laser cutter for cutting compound shapes, it was useful for rapid prototyping and quickly cutting repetitive patterns, saving manufacturing time.

IFMI is a "CAD to Master" facility — once the detailed design of a component was complete, the CAD models were directly imported from SolidWorks to most of the CNC machines. Amada's AP100US software automatically flattened and imported the SolidWorks models. The CNC

▲ Batman and Robin performed a three-part sequence during the hand-off of totes. Batman carried a recycling container to Robin, which had already accumulated a stack of six totes. Batman then docked with Robin, and control of the stack was transferred. Finally, Batman backed away to deliver the capped six-stack to the scoring platform.

operator analyzed the flat patterns to attach irregular holes that were not recognized by the software and defined the edges and attachment points. Once the patterns were programmed, Amada's AmNEST software was used to nest parts together to get the maximum number of usable parts from a single sheet of material, thus minimizing waste. The nested sheets were then cut into the desired pieces.

THE CNC ROUTER AND METAL BRAKES

Halfway through the 2015 season the team gained access to a CNC router, a three-axis computer-controlled cutting machine used for creating precision patterns in a variety of materials. Unlike the other shop machinery, both mentors and supervised students were allowed to operate this new tool. They used it to rapidly manufacture high fidelity prototypes out of cardboard, polycarbonate, and aluminum to verify calculations and assumptions before sending final models to the sheet metal shop.

The CNC metal brakes, used to bend the patterns cut from sheet metal, were the only equipment that required effort beyond the SolidWorks models. Minimalist drawings had to be generated that called out flange lengths, bend angles, and overall dimensions for the operators to program the machines.

Once the completed components were returned from the shop, assembly went quickly, as did component integration. All parts were modeled to such a precise level of detail that the physical components arrived from the shop exactly as designed. The schedule reduction from accelerated manufacturing and construction gave the team more time for prototyping, detailed design, and iteration.

ROBIN — THE STATIONARY STACKER

Robin's sole job was to stay parked in front of the loading station tote chute and build stacks of totes for Batman to retrieve. A single MiniCIM motor drove two continuous chain loops, each running parallel to a vertical track. Power was supplied to this motor via a tether from Batman, which held the control system, battery, power distribution, and motor controllers.

Two independent but synchronized sheet metal carriages traveled up and down the tracks on ball bearing rollers. Each carriage held a polycarbonate blade which deflected when pushed down over the lip of a tote, and then snapped back into place once past it. When this one-way tote gate closed, the carriage reversed direction to lift the tote. In this manner, Robin would pick up each of the six totes fed by the human player through the tote chute. Polycarbonate guides helped the totes fall into correct orientation when exiting the chute door and helped keep totes aligned during stacking to prevent jamming. Misalignment was also mitigated by coating the underside of Robin in a high traction latex material to maintain its position at the loading station. Robin's interior was lined with low friction plastic to help with stack extraction.

HOLY CNC-MACHINED SHEET METAL, BATMAN!

Batman manipulated stacks and grabbed recycling containers with a "touch it, own it" intake system — driver precision and robot alignment were not critical for acquisition.

The intake, which manipulated tote stacks and recycling containers, utilized chain-driven wheels. The outer wheels were made from neoprene drive rollers and the inner wheels were custom-designed and cut using a waterjet.

The two rear wheels of the robot's drivetrain were each directly driven by a CIM motor and custom two-stage gearbox.

Each side of the intake included two hard-mounted wheels chain-driven by a BAG motor with a VEXpro VersaPlanetary Gearbox. Sure-Grip neoprene drive rollers were used for the outer wheels and the inner wheels were made from custom-designed neoprene, waterjet-cut by FRC Team 118, the Robonauts. The soft surface and compressibility of the neoprene was advantageous when collecting irregular or misaligned game pieces and when pulling stacks out of Robin. Batman's tote stacking mechanism used carriages with polycarbonate blades similar to those on Robin.

Batman used a four-bar linkage arm outfitted with a pair of forks at the end for lifting recycling containers. The custom links were made from false extrusions — sheet metal C-channels riveted together to form a rectangular tube. The segments were laser cut to include bearing and mounting holes, as well as truss-inspired weight reduction patterns. The arm was direct-driven by two custom sheet metal spur gear reductions, powered by a BAG motor with a VEXpro VersaPlanetary Gearbox. The spur gears were made from multiple 0.125-inch-thick aluminum laser-cut gears that were stacked and riveted together, then directly attached to the arm.

The laser cutting capability not only enabled inclusion of mounting features and weight reduction patterns, but also allowed the team to only cut the part of the gear they needed. The arm only rotated 90 degrees, and thus didn't require 360 degrees of gear teeth. The upper tower, manufactured using parallel plate construction, supported the arm pivot points and gearbox. The recycling container arm and a secondary wishbone-shaped arm were also used to clear a path for Batman to pick up the yellow totes. Both arms were outfitted with special hooks designed to release the containers at the end of the autonomous period.

Two low-profile forks, fabricated from thin-walled aluminum tube, were attached to the end of the arm to pick up the recycling containers and place them on top of six-stacks of totes. The intake pulled containers onto the forks, which were angled down to facilitate acquisition. Prototype testing showed that performance wasn't affected by

The U-shaped chassis accommodated the intake, stacking mechanism, and tote stacks that were easily extracted from Robin.

> The rear drive wheels were connected to the front wheels with a chain-in-tube configuration, with the drive chains running inside the chassis rail tubes. Since the chain barely cleared the sheet metal above and below, it couldn't come off the sprocket unless it was purposely broken.

the horizontal or vertical position of the containers; they could be placed on top of stacks in the same orientation that they were picked up. As the forks were lifted, spring-loaded action tilted them up, increasing the stability of the recycling container and assisting placement on top of a stack of totes.

The primary design requirements placed on Batman's drivetrain were that it be lightweight, simple, and reliable so the team could devote weight margin and build time to the more complex robot features. The four-wheel drive system consisted of two rear six-inch VEXpro Omni Wheels that were each directly driven by a CIM motor and custom two-stage gearbox. These rear wheels were connected to two front six-inch VEXpro Traction Wheels utilizing a chain-in-tube configuration. The drive chains were run inside the chassis rail tubes with slight clearance between the sheet metal above and below, preventing the chain from accidentally coming off the sprocket.

The drivetrain components were mounted onto a reinforced U-shaped chassis designed to accommodate the intake, stacking mechanism, and totes. To preclude pool noodles from being pulled into the drivetrain, which the team experienced in early matches, a noodle fan was added. Made from a hobby remote control airplane propeller and BAG motor, it was capable of clearing a path up to six feet in front of Batman.

ACHIEVING THE GOAL

The Robowranglers value continuous improvement as one of their most important design principles. Throughout the build and competition seasons, the team was constantly refining and upgrading Batman and Robin's design to enhance performance on the field. The CNC rapid production capabilities available to the Robowranglers offered the team a way to quickly modify parts or replace entire subsystems.

Considering that the team's high-level goal was to win the *FIRST* Championship by maximizing the number of points the robot could score, it came incredibly close. FRC Team 148 was the top seeded team and finalist at the Dallas Regional; the top seeded team and winner of the Las Vegas Regional; and the top seeded team and winner of the *FIRST* Championship Curie Subdivision. FRC Team 148 also played up through the semi-finals on the Einstein field. The Robowranglers were one of the highest scoring FRC teams in 2015, with Batman and Robin reliably able to build at least three capped stacks of six totes in every match. There is no arguing that they achieved their objective.

FRC Team 148 is proud of the successful process it used to analyze the game, choose a strategy, develop robot concepts, and eventually build Batman and Robin. The team's proven skill in sheet metal design and manufacturing empowered it to create two extremely successful robots. As FRC Team 148 continues to expand its knowledge and use of CNC capabilities, it will be better prepared for the 2016 challenge. To see what the Robowranglers improve upon, tune in next season — same Bat-time, same Bat-channel.

> Similar to Robin, Batman incorporated custom sheet metal carriages to stack yellow totes during the autonomous period. The carriages, arm components, and gears were all fabricated with custom mounting features and weight reduction patterns.

Team 1983 – Above and Beyond with CNC Cutting Capability

CUSTOMIZED WEIGHT REDUCTION WITH WATERJET MACHINING

FIRST® Robotics Competition (FRC®) Team 1983, Skunk Works Robotics, knows how to use computer numerical control (CNC) cutting tools to build an elegant and competitive robot. Its unique team name was inspired by sponsor Lockheed Martin's Advanced Development Programs, also known as the Skunk Works®. This secret program was responsible for many revolutionary aircraft, including the SR-71 Blackbird and the U-2 high altitude reconnaissance aircraft.

Surrounded by Seattle's aviation-related businesses, Raisbeck Aviation High School is preparing students to be the next generation of aviation and aerospace-focused scientists and engineers. This team put its strategic planning and innovation skills to the test by designing and building the majority of its robot with waterjet-machined components.

REDUCING RISK WITH RAPID PROTOTYPING AND MASTER GEOMETRY

The previous prototyping methodology employed by FRC Team 1983 involved narrowing down to one or two ideas, building mockups, and then modifying the mockups until they worked. This process was slow and ineffective and was not producing successful robot designs. To improve design and test as many ideas as possible to find the right solution, the team realized it would need to overhaul its prototyping process.

For the 2015 season, Skunk Works Robotics adhered to an established rapid prototyping program, developed by team members, to best utilize the knowledge base of the team as a whole in the initial period following the season's start. The high-level problem to be solved was how to utilize a systems approach to select the best robot design as soon as possible that would be successful in competition. The solution began with a thorough understanding of the game and robot requirements and identification of a strategy. For each desired robot function, subsystems were selected to move on to prototyping. The team then separated into smaller groups to build the prototypes — each team only had one week to prove the feasibility of the selected solutions. The prototyped systems were then discussed and compared, with reliability and feasibility heavily weighed to determine the final robot design.

The final step of the rapid prototyping program was to create a rough two-dimensional (2D) representation of the imagined robot in Autodesk Inventor, referred to as the master geometry. The scaled sketches included major robot components, and each part was designed to reference the master geometry so that if one part was modified, the relative position of the other components would automatically be updated.

These sketches helped the team visualize potential design limitations and identify critical interfaces of all

Rapid prototyping and a systems-based approach were used by FRC Team 1983 to refine initial concepts, resulting in a highly competitive robot.

A MAXIEM 1515 waterjet was used to fabricate both prototypes and final components for the robot. Each sheet metal part was modeled using Autodesk Inventor software, then converted to a two-dimensional drawing and imported to special software that created toolpaths for the waterjet.

subsystems. Constructing a whole-robot view in design software helped the team anticipate interferences, identify structural flaws, and estimate the robot's weight before proceeding to detailed component design.

ABRASIVE WATERJET MACHINING

Most of the custom fabricated parts on the robot were machined using a MAXIEM 1515 waterjet. This method of cutting can be used on a wide variety of materials, including plastics, steel, alloys, glass, and stone. It is especially useful when cutting materials that are sensitive to the high temperatures generated by other machining processes; the risk of changing material properties is eliminated without a heat affected zone. Pressurized water is pumped through a nozzle where it mixes with a granular, abrasive material such as garnet — a hard gemstone with sharp edges when fractured. When the high speed water and abrasive mixture exits the nozzle, it cuts through the material positioned below.

Since the MAXIEM 1515 can accommodate stock material up to five feet wide by five feet long, the team was able to use this tool for cutting wood prototypes, and also aluminum, steel, and polycarbonate panels for the final robot. The students generated a model of every part on the robot in Autodesk Inventor and then created separate drawings of the robot's sheet metal parts. The drawings were imported to OMAX Intelli-MAX LAYOUT, computer-aided design (CAD) software that creates toolpaths for the MAXIEM waterjet, and the mentors operated the machine to create rapid prototypes and final component fabrication.

FRC Team 1983 began using a waterjet for robot fabrication in 2007, bringing parts that required machining to the facility of waterjet-manufacturer OMAX. The relationship between OMAX and the team grew quickly. OMAX employees joined the team as programming and fabrication mentors, and team members were offered internships with OMAX.

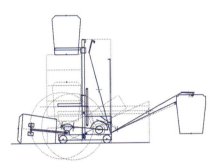

The robot's master geometry, created using Autodesk Inventor, was used to visualize the relative placement of components and applicable size constraints.

A model of the final robot assembly illustrates the integration of all major subsystems.

The waterjet, used to fabricate aluminum chain connectors on the elevator, enabled the inclusion of threaded holes for tensioning bolts.

The tote stacker assembly was carefully designed to accommodate the weight of the totes and to prevent misalignment on the vertical track.

Angled polycarbonate blades attached to the tote stacking arms deflected when lowered, then snapped into place to securely lift each tote.

The elevator was driven by two motors mounted to the vertical track supports. Autodesk Inventor Design Accelerator software was used to calculate the parametric values of the spur gears and chains for the system.

The partnership with OMAX has helped the team build increasingly sophisticated robots with shorter manufacturing time. In 2013, the team purchased its own waterjet from OMAX for a new prototyping lab at the school. This added flexibility to the team's operations; the team's access was no longer dependent on availability of the OMAX facility. Of all of the equipment available to the team, the waterjet was used most extensively.

FLANGES FOR JOINING COMPONENTS

To attach metal plates at right angles, the team had previously used 0.75-inch aluminum angle, attaching each plate to the flanges. In 2013, they acquired a box and pan brake, which introduced the ability to accurately bend flanges in sheet metal. The brake clamps onto the metal piece to become a flange and bends it to the desired angle. This tool complemented the waterjet capability because sheet metal pieces could be designed with tabs so that when bent they would become flanges to be riveted to other sheet metal parts. While this ability accelerates manufacturing, special consideration must be taken when designing the shape of the sheet metal. The process of how each bend will be made in the brake, as well as avoiding interference with clamping teeth, must be planned into the design. To standardize their process, the team members used a common 0.75-inch size for all flanges, reducing machine calibration time.

DUAL-PURPOSE ELEVATOR

The final robot was designed with an elevator device that served a dual purpose: to stack totes, as well as lift and stabilize recycling containers. Up to six totes were stacked by two parallel arms with angled polycarbonate blades machined using the waterjet, which snapped under the lip as they descended over each tote. These arms also had the ability to extend forward for depositing yellow totes on the step for extra points. This stacker assembly rode on a vertical track, which was shared with a recycling container stabilizer mechanism riding directly above. The container stabilizer consisted of a pneumatically actuated claw that was designed to grab the base of a recycling container. The material for the symmetric claw components and the waterjet pattern for weight reduction were selected specifically to prevent bending while under force. Wheels were attached to the ends of the claw to help guide the containers into a secure position and provide three points of contact to prevent the containers from slipping during transport and while totes were stacked below.

The elevator frame was made of three welded aluminum C-channels that held Delrin rollers for the moving stacker and container stabilizer. This system was large and contributed significant weight to the robot's allowance. Most of the components were cut using the waterjet to add accurate mount holes, but also to reduce weight. A pattern was created for the waterjet to remove as much material as possible from the structural components, while maintaining the required strength and rigidity.

EXTENSIVE USE OF A WATERJET FOR ROBOT COMPONENTS

Pivoting arms extended out in front of the robot's stacker arms and claw. The arm ends were outfitted with articulating wrists and dual mecanum wheels chain-driven by RS550 motors. When the spinning wheels made contact with a horizontal recycling container, they aligned the container and pulled it closer to the chassis. The arms then pivoted back to a vertical orientation, which positioned the claw to grip the base of the container. The robot was designed to pick up recycling containers from a horizontal, or knocked over, position because the team's analysis revealed that the containers were likely to be knocked over during matches, as they were more stable when on their sides. If recycling containers weren't already knocked over, the robot would do so to position them correctly for acquisition. The waterjet enables parts to be fabricated with tighter tolerances than if handmade, which was important for this system to function properly. To keep the chain taut and ensure gears meshed properly, the waterjet was used to machine a plate that referenced the motor with two hard stops, thus the gear and chain attachment points were exact.

A pneumatically actuated claw was used to grasp recycling containers and stabilize them during transport.

Wheels at the ends of the claw guided the recycling containers into position and provided a secure grip when the claw closed.

A full model of the dual purpose elevator assembly helped determine how the separate components fit together.

Totes could be fed from the loading station through the back of the robot onto a polycarbonate ramp.

Computer-Controlled Cutting Systems | 191

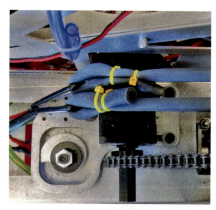

A custom plate was made using the waterjet to adjust the chain tension and maintain alignment between the motor and gears in each recycling container arm.

Pontoons were attached to the sides of the U-shaped chassis frame to add rigidity and prevent flexing while manipulating tote stacks.

Motor-driven mecanum wheels, mounted to the ends of pivoting arms, pulled recycling containers into the chassis.

A U-shaped chassis allowed the intake of totes from the landfill zone through the front of the robot. The chassis frame was designed to be as lightweight as possible while still exhibiting sufficient structural integrity to withstand impacts. To reinforce the U-shaped chassis and prevent flexing, especially when the robot was carrying heavy tote stacks over the scoring platform, strength and rigidity were increased with the addition of pontoons along the sides of the chassis frame. The drivetrain consisted of four individually powered mecanum wheels, arranged in a square shape, for omnidirectional maneuvering.

The robot was also designed to receive totes from the loading station, being fed behind the elevator structure. A ramp made of clear polycarbonate sheet cut by the waterjet was installed above the drive base to correctly position the totes for the stacker.

Once the majority of the robot had been completed and the team members knew they had some weight margin, they decided to add a recycling container grabber for use during the autonomous period. Mechanisms inspired by tape measures and umbrellas were considered, but the team determined these would not meet speed and reliability requirements. The selected solution, inspired by another team, consisted of a prong with five bolts that slammed onto the lid of a recycling container, then was dragged backwards, catching the bolts in the center hole of the lid.

The efficiency of the extensive material removal for weight reduction would not have been possible without access to CNC machining. The waterjet could accurately and rapidly machine the diamond-shaped truss seen throughout the robot's structural components. The waterjet capability enabled this team to change its entire design and build processes, and easy access meant the team could cut parts whenever and as often as needed.

According to FRC Team 1983's website, one of its objectives is to "design and build high quality products with defined purpose and pride." The team's 2015 robot clearly shows that it has achieved this objective. Winning all three of its district events, as well as the Pacific Northwest District Championship, there is no doubt that these aerospace-inspired students produced an exceptional robot that went above and beyond.

The majority of the robot was designed to incorporate waterjet-cut components. The speed, accuracy, and flexibility to support a variety of materials made waterjet cutting an integral part of the team's design process.

Team 2848 – The Right Tools Enable Flexibility

EMBRACING DESIGN EVOLUTION

As winners of the 2014 *FIRST®* Robotics Competition (FRC®) Championship, Team 2848 — the All Sparks — have experienced what it takes to build a winning robot. This Dallas, TX, team channeled that success to focus on continuous improvement. The All Sparks pride themselves on giving students hands-on experience with modern machining tools and industry practices. The team's strategic use of computer numerical control (CNC) laser and plasma cutters enabled the machining flexibility to continually enhance the robot throughout the competition season.

MODERN MANUFACTURING TOOLS

The All Sparks traditionally construct prototype components from plywood and two by fours — accessible, low-cost materials that are relatively easy to work with. With lessons learned from past robot prototypes, the team members found that to make an accurate final robot component based on the specifications of the prototype, the accuracy of the prototype's dimensions and tolerances is imperative. They learned to anticipate different mechanical properties of the wood compared to metal components, especially flexing of materials. While the wooden mockups could be built rapidly, they could also slow down the design process if the team had

◂ *The All Sparks used computer numerical control laser and plasma cutters to fabricate robot components. The laser cutter was used to cut medium density fiberboard for prototyping of the robot's claw.*

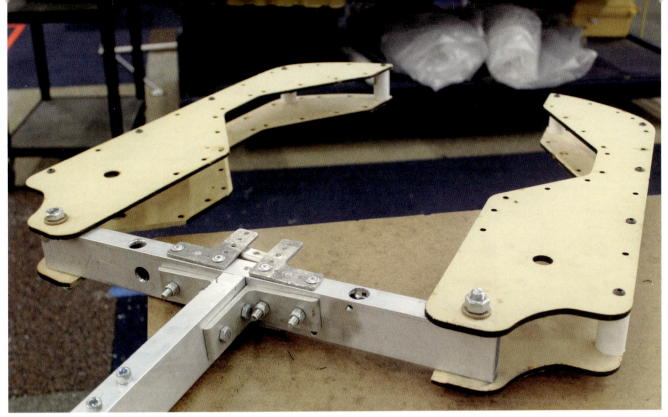

◔ While most of the final robot components were fabricated from sheet metal, less expensive yet fully functional wooden mockups with pressed-in bushings and bearings were used for proof of concept testing.

to use valuable time troubleshooting and fixing discrepancies to correct for the differences in performance.

Prior to the 2015 competition season, the team had access to a small CNC mill and a CNC plasma cutter. To accelerate the fabrication process and reduce uncertainties in prototype accuracy, the team chose to invest in additional equipment. With the good fortune of an available budget, FRC Team 2848 was able to purchase new tools for their workspace. To better cut wood and plastic, the team invested in a carbon dioxide (CO_2) laser cutter. This tool directs a focused beam of infrared light, invisible to the human eye, through a nozzle. Carbon dioxide is also fed into the nozzle, and the exiting laser beam and oxygen combination vaporizes or melts the material being cut. The team also purchased two PolyPrinter three-dimensional (3D) printers, a hydraulic brake press for bending sheet metal, a manual milling machine, and various power tool accessories.

The All Sparks' robots are typically fabricated with sheet metal and these new tools significantly sped up the manufacturing process. Students could model the robot components using SolidWorks, then upload them to the laser cutter to machine a wooden prototype. Bearings and bushings could be pressed into the mockup to create a fully functional prototype built to the identical tolerances as the final metal parts. After testing, any required changes to the prototype were modeled in computer-aided design (CAD) software, and, with minimal modification, these same electronic design files were uploaded to the plasma cutter to fabricate the final metal parts.

◔ Laser-cut wooden prototypes were mounted to the drivetrain to determine how the components would interact.

▲ The first iteration of the competition robot incorporated a recycling container claw attached to a four-bar linkage. FRC Team 2848 streamlined the decision making process by establishing a design framework and standards. These guidelines accelerated the robot build and assembly processes.

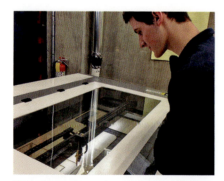

▲ The laser cutter used a combination of carbon dioxide and a focused beam of infrared light to vaporize the base material. A student monitored the machine as one of the medium density fiberboard prototype components was cut

ESTABLISHING A FRAMEWORK TO ACCELERATE DESIGN

With only six weeks to design and build a robot, teams are compelled to find innovative methods for streamlining the process. Efficiency improvements can often be found in the design decision methodology and, as a result, fabrication of the robot. FRC Team 2848 discovered that it could accelerate the process by establishing a design framework consisting of a set of standards to adhere to. A design framework identifies different strategies to help guide wise decision-making. This team has been building robots since 2009, thus had several years of experience to reference for developing these standards. Team members analyzed best practices and successes and failures, and identified a list of components that could benefit from an established set of guidelines governing build and assembly processes.

The team identified six standards to assist the students with making quick design decisions. All rotating axles on the robot were fabricated from 0.5-inch outer diameter, 0.25-inch inner diameter aluminum tubing. This material proved rugged enough to withstand the rigor of competition and could easily bolt to connecting components through the hollow center. All bearings rotating at a speed of 500 RPM or greater had an inner diameter of 0.5 inches and an outer diameter of 1.125 inches, a size easily joined with other standardized components. For slow spinning and low load rotating systems, all bushings were bronze with a 0.5-inch inner diameter and a 0.625-inch outer diameter. For coupling components to

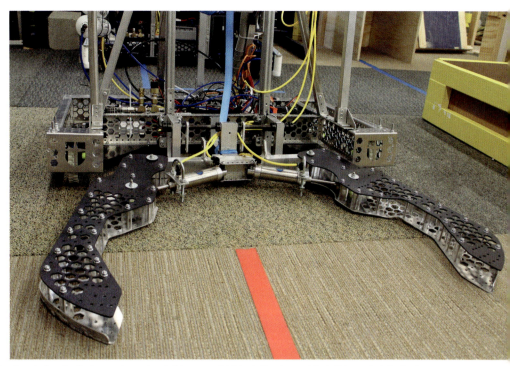

▲ The recycling container claw went through multiple iterations before the final geometry and materials were established. The shape of the initial rounded plastic plates was modified to better grip the containers, and the bottom plastic plate was replaced with metal for added strength.

a spinning shaft, a 0.5-inch hexagonal shaft was used. To simplify design, size 10-32 standard hardware was used throughout the robot to reduce the number of tools required for assembly and maintenance, including the universal use of a #9 drill bit for all clearance holes. Finally, if parts needed to be connected but bolting wasn't preferred, 0.1875-inch pop rivets were used. This size of rivet was chosen because the required hole size matched that of the standard hardware.

The efficiency and effectiveness of the design process was also improved through development of student CAD skills. In the months leading up to the 2015 FRC season, all students on the team were encouraged to participate in an online SolidWorks training program. The All Sparks recognize that while not every student needs to be proficient in CAD, it is helpful for team members to have a basic understanding of the robot design software. This independent learning program provided a foundation for students to communicate their designs more effectively.

ADDING CAPABILITY TO IMPROVE PERFORMANCE

The initial robot design incorporated a laser-cut claw attached to a four-bar linkage. The upper and lower plates of the claw were constructed from two pieces of acrylonitrile butadiene styrene (ABS) plastic sheet and were offset by threaded rod and acrylic spacers to give the claw more gripping surface. The silhouette of the claw evolved several times to find the best shape for gripping recycling containers. These rapid revisions were enabled through use of CAD and laser cutting multiple prototypes for testing. Between the claw and linkage was a pivoting, wrist-like joint that allowed the robot to pick up the recycling containers from any orientation and return them to an upright position for capping stacks of totes.

The first regional competition attended by the team was an opportunity to see how well the robot interacted with alliance partners. Analysis of the design and functionality prompted the team to make several improvements before its next competition. While the initial strategy focused on manipulation of recycling containers, the team decided to add a tote stacking capability to increase the robot's competitiveness. The four-bar linkage system was replaced with an elevator-style lift. The

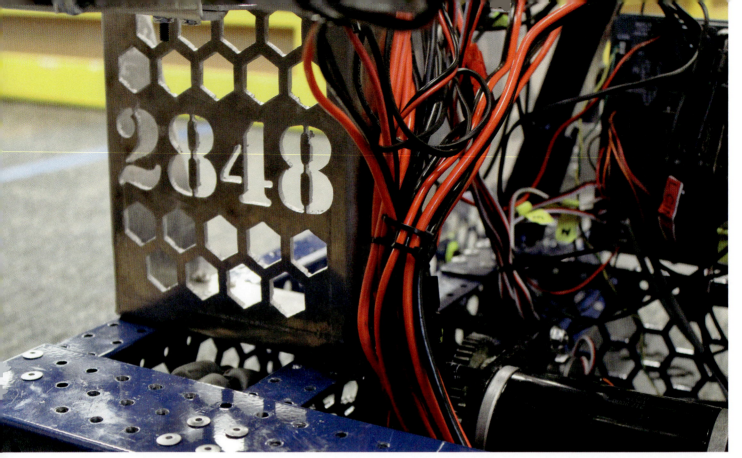

⬢ The team's Hypertherm plasma cutter used pressurized gas and an electrical arc to melt the material being cut. This tool enabled rapid fabrication and iteration of metal plates for the final robot, which were customized when possible with a cutout of the team number.

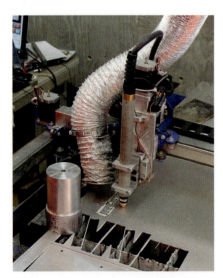

⬢ The high heat produced by the plasma cutter can cause thin materials such as aluminum to warp. To minimize deformation, the team used other heavy materials to hold the aluminum down, keeping it flat.

new design enabled the robot to lift game pieces vertically, which improved weight distribution and stability.

Minor modifications were also made to the claw, including a revised shape to grip totes as well as recycling containers. The bottom plastic piece was replaced with aluminum plate for additional strength. The plate was manufactured using the CNC plasma cutter. Similar to a laser cutter, pressurized gas is forced through a nozzle at high speed. An electrical arc is generated between the nozzle and the material being cut, and as the gas passes through this arc it is heated until it turns to plasma. This stream of plasma melts the metal, thus cutting the material.

ADAPTING TO MAINTAIN COMPETITIVE ADVANTAGE

FRC Team 2848 chose to focus on a strategy that involved manipulating the recycling containers. It anticipated that the ability to control these scoring multipliers would be a crucial asset at the higher levels of the game. To be successful at this, the team's robot underwent several design evolutions throughout the season.

To maximize the number of recycling containers an alliance could score, it was imperative for the alliance to gain control of as many recycling containers from the center step of the field as possible. To achieve this, one student took on the initiative to

○ The initial design of the mechanism that grabbed all four recycling containers off the step consisted of telescoping arms. At the end of the autonomous period, the arms were retracted with two motors that reeled in string attached to the ends of the arms. At full extension, the arms spanned a width of over 20 feet.

design a mechanism that could grab all four center recycling containers at the start of each match. Because this device was only briefly required at the start of the autonomous period, it was designed to retract so as not to interfere with the other robot functions during the remainder of the match. The device consisted of two fold-down telescoping poles, affectionately dubbed the "wings of doom," mounted to an actuator that extended them away from the robot.

The retraction of the telescoping components was enabled by two small motors to reel in high-strength string attached to the ends of the 20-foot wingspan. At the end of each telescoping pole section, the team attached a set of slightly curved spring-loaded fingers fabricated on the laser cutter. Upon extension of the poles, the plastic fingers were hinged in such a way as to fold in when suspended over the containers. This design caused the fingers to slide over the top of the recycling containers until they reached the hole located in the center of the lid, upon which they extended into a downward position. With the fingers grabbing all four recycling containers, the robot retreated toward the driver station, pulling the recycling containers with it.

As the competition season progressed, more and more grabbing mechanisms were appearing on other robots. The All Sparks' robot lacked the speed to grab all four recycling containers before or as quickly as opposing alliance robots

○ The laser cutter was used to fabricate ABS plastic gears for a prototype. This material was used throughout the robot due to its strength, flexibility, and ease of manufacture.

▲ As other robots developed increasingly faster mechanisms to grab recycling containers off the step, FRC Team 2848 continued to evolve its system to stay competitive. The final revision was smaller and faster than the original design. Two carbon poles were extended over the containers, and tethered wiffle balls became snagged in the container lid openings.

▲ Spring-loaded plastic fingers were attached to the ends of the telescoping arms to grab the recycling containers through the holes in the lids.

could grab fewer containers. The team decided to replace its four-container grabbing mechanism with a smaller, more lightweight adaptation that could quickly grab the two center containers.

The system evolved yet again for the FIRST Championship. Other teams had developed extremely fast, whip-style grabbing mechanisms — speed was now critical to take advantage of the score multipliers. FRC Team 2848's final iteration looked very different from its initial design, but provided the speed needed to successfully gain control of any set of two adjacent recycling containers. The final design consisted of two motor-driven, carbon fiber poles, each with a string-tethered hard plastic wiffle ball on the end, wrapped in tape to customize its shape. The poles were extended such that the balls became snagged in the recycling container lid opening, and again the robot dragged the containers onto its alliance's side of the field.

FRC Team 2848 members understand the hard work required for success, and they eagerly learn from their experiences to improve both the team and their robots. By establishing a design framework, team members continue to streamline their processes while setting the stage for a smart robot. The team's knowledge and application of machining processes were evident in the 2015 robot's evolving design. More important than being equipped with the right tooling, this team equipped its students with critical thinking skills and open minds — traits that will continue to support the All Sparks' evolution.

FRC Team 2848's robot continued to evolve throughout the build and competition seasons. The use of laser and plasma cutters provided the manufacturing flexibility needed for this continuous improvement.

Team 4183 – Wood 'bot, Good 'bot

GOING AGAINST THE GRAIN

A community robotics team is one where the members are from the community at large, as opposed to a specific high school. One of the challenges of community teams is finding a place where the team can meet and work. *FIRST*® Robotics Competition (FRC®) Team 4183, the Bit Buckets, is a community team housed in the Xeocraft Hackerspace in Tucson, AZ. Not only does the home provide access to the tools to build a robot, it also infuses the team into an existing community of makers and hackers.

This 2015 robotics team was motivated to pursue a unique design concept: building a robot using wood as the primary construction material. As a team entering its fifth year, FRC Team 4183 defied the conventional FRC thinking that wood is for beginners and working in metal is better for experienced teams. By defying conventional thinking, the team proved that advanced technologies can be applied using wood and that wood construction can be just as effective as metal construction.

TESTING THE IDEA

To construct a wooden robot the team capitalized on available opportunities, including access to technology and the creativity of its maker community. The

❯ *Laser-cut birch plywood was the primary structural material on this unique robot. Two-dimensional shapes were joined together to create three-dimensional components, which were, in turn, connected to other components to build a functional robot.*

◉ A tab and slot construction method was used to build box elements, with fasteners screwing into embedded plates to hold the pieces together.

◉ The construction of a drivetrain was the team's first foray into building components using laser-cut plywood. That drive train was one component in a prototype of the robot's base. The prototype's strength was evaluated by subjecting the composite structure to a rather unconventional load.

team also capitalized on the inherent structural properties of wood as a fabrication material. Within the Xeocraft Hackerspace, the team had access to a Trotec Speedy 300 laser cutter that was capable of cutting 0.25-inch-thick pieces of plywood. The maximum-sized sheets that could be cut by this laser cutter measured 29 inches by 17 inches, thereby prompting a need to design modules that could be combined to create larger structures.

Baltic birch plywood was selected as the building material, based on its uniform quality, superior structural properties, and cleanly finished appearance. A modular approach for creating and joining box structures was adopted, capitalizing on the inherent triangulation of a rigid box that resists torsion and supports substantial compression and bending forces.

A clever mechanism for attaching individual pieces of laser-cut wood as well as modular sections was made up of laser-cut tabs and slots, which were held together with screws and captive bolts. Iteration identified the optimum sizes for the tabs and slots that firmly held each piece and its fastener. A horizontal slot was cut into each vertical slot, with a square nut then embedded into the horizontal slot. A screw, passing through a hole in the attached tab and through the vertical slot to the embedded nut, secured the two laser-cut pieces together. This same technique was used to attach individual sections and create large components.

This construction method was evaluated prior to the start of the build season by constructing a laser-cut drive base composed of modular sections. In this test case, each side of the base included a detachable drivetrain to examine the construction method's ability to maintain the tolerances needed for geared systems. The validity of the design methodology's strength was evaluated in a whimsical fashion by subjecting the drive base platform to an artificially high load, in this case provided by a 1965 Volvo that was driven over the platform. The platform survived the near-destructive testing, increasing the team's confidence in this construction approach.

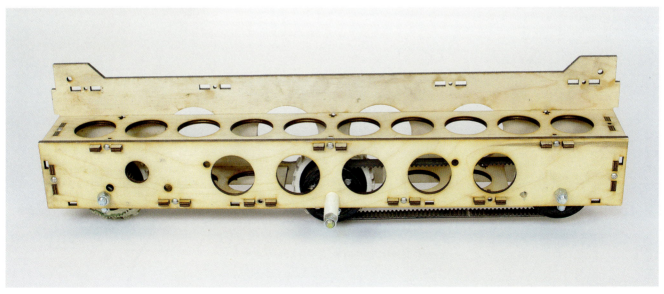

A Trotec Speedy 300 laser cutter provided up to 120 watts of power for cutting material. The laser was capable of cutting 29-inch by 17-inch wood panels having a maximum thickness of 0.25 inches.

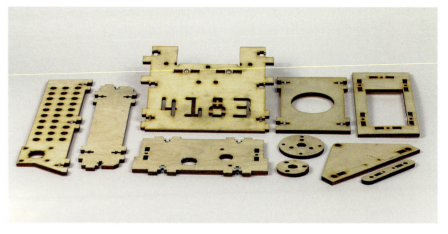

A modular approach created boxes out of laser-cut plywood. The overall structure was subdivided into multiple compartments, with each compartment a rectangular box having interlocking sides.

A CAD rendering included all of the laser-cut structural components. Versions of the CAD files were used in the laser cutter to fabricate the plywood profiles that were then combined to create rectangular structures.

STRATEGY AND DESIGN PROCESS

The Bit Buckets established three strategies to compete in the 2015 FRC competition RECYCLE RUSH℠. The robot would harvest totes from the landfill, stack a recycling container as high as possible, and have a drivetrain with a high degree of maneuverability. After establishing these strategies, a list of functional and performance requirements was crafted to guide the design process for creating the robot.

The design process began with small groups sketching mechanisms that achieved the strategies: a tote lift system, a recycling container manipulator, and the drive base. Promising ideas were developed as simple prototypes using wood and aluminum. Laser-cut plywood was only used in the prototyping phase if dimensions were critical to evaluate the design. Concepts were modified until they achieved the desired performance for acquiring, lifting, and maneuvering game pieces. During this phase all sketches and results were documented in an engineering journal maintained on a remotely accessible website — a technique that made the information easy to share among all members of the team.

Once suitable designs for each system were identified, computer-aided design (CAD), using SolidWorks, was initiated. Two-dimensional (2D) models were first created to evaluate the fit, balance, and reach of each mechanism. Animation was used to verify the operation of each component, followed by three-dimensional (3D) modeling or prototype testing of individual mechanisms to validate assumptions. The design progressed to a 3D concept of the complete robot, with this model then amended with details such as the sections, slots, tabs, and holes needed to hold the individual pieces together.

The complete process consisted of a progression from the original sketches to the detailed construction plans that were exported to the laser cutter for manufacturing. The individual components were then assembled into modules, with the modules combined to create subsystems that were, in turn, joined to create the functional robot. The process was applied to first create an initial prototype of

⬤ Team members assembled the chassis using the box construction method. Three plywood sections were layered to form the bottom of the robot, with vertical cross members providing rigidity and strength.

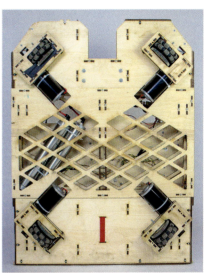

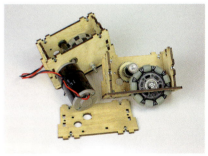

⬤ Each structural member of the robot included many parts that were fastened together. The drive transmissions were also assembled using this manufacturing methodology as the laser could produce shapes with the tolerances needed to properly align gears.

the complete robot, followed by an iteration cycle where improvements were incorporated and the final design perfected. The use of laser-cut parts eased the ability to make changes as well as manufacture spare parts.

The CAD document management software GrabCAD Workbench was used to manage and share the SolidWorks CAD files. Using this system, all team members could synchronize their individual contributions with the latest design. The software was also used to create a standard design library of field elements, commercial off-the-shelf parts, and custom plywood design features. Plywood elements including bearing bores, dovetails, tabs, and slots were created and saved in the design library as SolidWorks sketch blocks. This allowed the features to be simply dragged and dropped into any drawing, speeding the design cycle time for each component.

BUILDING A 'BOT WITH PLYWOOD

An omnidirectional drivetrain was designed and constructed using the laser cutter. As with other components, individual pieces were assembled to construct the drivetrain and the robot's base. Each corner of the robot base housed a four-inch omnidirectional wheel powered by a CIM motor. The motor, gears, and wheel were independent modules that attached to the drive platform at each corner, mounted at 45-degree angles to achieve omnidirectional motion when driven.

The laser-cut motor pods provided the needed accuracy for mounting these precision parts and were sufficiently strong enough to withstand the competition demands. A removable electronics panel mounted above the drive base provided ready access to the electronics. The base of the drive system served as a mounting location for the battery and compressor. This mounting position and heavy weight of these components helped establish the robot's low center of gravity.

The tote stacker evolved from a carriage rolling on aluminum rods to a C-shaped tower constructed out of birch plywood. Testing showed that the C-shaped configuration became twisted due to torsion, a condition that persisted even when the tower was

◆ The curved section of the stacker provided the robot with an ability to pick up totes when the stacker was fully tilted back.

◆ The claw and carriage picked up containers in any orientation and placed containers on stacks up to five totes tall.

◆ A tape measure was the primary mechanism for the container-grabbing feature. This device was designed in partnership with FRC Team 842.

reinforced with additional supporting structure. Tapping into prior experience, the stacker was redesigned as a three-inch by four-inch wooden tube, a solid and stable configuration used throughout the robot.

A carriage was equipped with hooks to grab the edges of totes and included bearings that rolled on the edge of the stacker plates. The bottom of the stacker column was vertical at its base to allow the lifting hooks to easily catch the tote edges. The stacker column then angled back into the robot to more securely carry each suspended tote. When the robot reached the scoring platform, the stacker column rotated to deposit the totes as a vertical stack. A winch powered by a motor located in the drive base provided the lifting force for this system. Laser-cut spools served as the winch's take-up reel, a design feature that allowed the torque and speed of the lift to be easily altered by simply cutting and installing different-sized spools. The final design was effective and robust, the result of an accelerated iteration process from the team's incorporation of laser-cut components.

The grasping and lifting mechanism to acquire the recycling containers consisted of a pair of arms mounted to a carriage. The arms relied on surgical tubing and a worm-geared, motor-powered transmission to open and close. The carriage rode up and down on one-inch square aluminum tubing, with these members being the only non-wood structural components on the robot. A series of skateboard bearings was mounted inside the carriage that supported this grasping mechanism and prevented binding. As with the ttote lift system, a winch provided the lifting force for the container-grabbing mechanism. The powered spools were fabricated on the laser cutter which allowed for easy tuning of the spool diameter to achieve the desired lifting force and speed.

A mechanism to quickly grab recycling containers from the center of the field — referred to as a "can burglar" with a nod to a once-popular fast food marketing icon — was designed in partnership with FRC Team 842 – Falcon Robotics – from nearby Phoenix, AZ. The designed device consisted of a motor-driven tape measure with a hook attached to the end of the tape. The system deployed in 0.25 seconds and was designed as a self-contained mechanism that could be added to both teams' robots. This mechanism was designed and manufactured following the team's first competition of the season. The team's nimbleness with laser-cut manufacturing minimized the time needed to produce a viable solution.

OPTIMIZING THE TECHNOLOGY AT HAND

The team's engineering strategy was to capitalize on the resources and facilities at hand. Those resources included the creative abilities of the maker community that comprised FRC Team 4183 and the laser-cutting technology that existed at the team's manufacturing site. The Bit Buckets' off-season development process to explore the feasibility for using laser-cut plywood as a construction material was an important aspect of the team's success. As a result of that exploration, the team produced a methodology to fabricate laser-cut wooden pieces that could be assembled into strong and durable robot components.

Moving into a new technology area such as building an entire robot out of laser-cut birch plywood required that the team was well prepared, that designs were carefully developed, and that all assumptions were fully tested. And in the end, the team's performance and the robot's endurance demonstrated the validity of applying this new technology for robot design and construction.

◆ Each tote was lifted by a separate carriage. A rope linked all of the carriages together to create an elevator. The laser-cut, winch-powered elevator pivoted forward to deposit totes on the scoring platform.

WOODIE FLOWERS AWARD

Model Mentor: Mark Lawrence, FRC Team 1816

The Woodie Flowers Award was established in 1996 by Dr. William Murphy to recognize outstanding *FIRST*® Robotics Competition (FRC®) mentors who lead, inspire, teach, and empower their teams using effective communication in the art and science of engineering design. FRC students may nominate one mentor from their team each year to be considered for the prestigious award. The recipient of the 2015 Woodie Flowers Award was Mark Lawrence from FRC Team 1816, The Green Machine, located in Edina, MN.

In 2005, shortly after retiring from a career at Cisco Systems, Inc., Mark was asked to help a pair of Edina students start an FRC team with a grant from NASA. FRC Team 1816 was born, and so began Mark's commitment to all things *FIRST*.

On The Green Machine's website, Mark is described as a believer that *FIRST* is not about the robot; it is about the students growing and pushing themselves to excel. He has high expectations for his team, and encourages both the students and mentors to achieve excellence. Mark's communication style shows his commitment to the students — he prioritizes their needs over all others. He helps students to identify their passion, readily assisting both his and other team's members with career planning. Mark's commitment to his team was recognized when he received the Connecting with Kids Commendation from the Mayor of Edina.

Besides being a lead mentor for FRC Team 1816, Mark is also driving the growth of Minnesota's *FIRST* program as the founding Chairman of the Minnesota *FIRST* Regional Planning Committee. Under his guidance, the number of Minnesota teams multiplied from two in 2005 to 192 in 2015, and the state now hosts four regional competitions with over 60 teams participating in each. Minnesota has the most FRC teams per capita in the United States: there are more robotics teams than there are varsity boys' hockey teams. Currently, 56% of Minnesota high school students have access to FRC programs, but Mark's aspiration is to make *FIRST* accessible to all high school students in the state.

Mark led a collaborative effort between Minnesota *FIRST* and the Minnesota State High School League to make FRC a fully sanctioned varsity high school sport. In 2012, the inaugural Minnesota State Robotics Championship became the first FRC competition of its kind in the nation to be sponsored by a high school activities association. Minnesota robotics teams now receive the same recognition as sports teams, and Mark believes that this not only fortifies FRC teams as part of the education system, but also transforms the way FIRST students are perceived by their peers. Mark is continuing the effort, assisting Arizona and Washington to achieve similar varsity sport recognition.

Mark understands the importance of staying ahead of the innovation curve and the need for scientists and engineers in Minnesota's future.

He promotes FIRST through his participation on the Edina District Technology Taskforce; as an advisory board member for Robotics Alley, an initiative to expand robotics and automation in the Midwest; and as a member of the University of Minnesota Computer Science Associates, lending his expertise and providing input on industry trends to the Computer Science and Engineering department. He encourages alumni of FRC Team 1816 to stay involved with FIRST, and their presence is prevalent as volunteers at competitions.

Mark inspires The Green Machine throughout the robot production cycle and beyond as he encourages the development of the future leaders on the team. His pride in being a member of FRC Team 1816 can be clearly seen at the competitions as he congratulates all teams on their accomplishments. Mark is described as the team's most fervent and enthusiastic cheerleader. He is not only a leader within the team, but for the entire state, as his passion for FIRST continues to drive growth in Minnesota. Mark's ambition for the futures of his students and FIRST is contagious, and the students can't wait to see where he takes them next.

> Surrounded by past winners, Mark Lawrence (second from the left) was presented with the 2015 Woodie Flowers Award at the FIRST Championship. Nominated by his team, Mark was recognized for decades of service inspiring and motivating FIRST students.

CHAPTER 7
SENSORS, MONITORING, AND CONTROL APPLICATIONS

A robot's control system receives input sensor signals, analyzes the data using programmed instructions, and actuates systems on the robot. The input signals can be initiated by an operator or generated by the robot control system based on programmed algorithms. Sensors are commonly used in control systems to monitor physical parameters and provide feedback. Programmed instructions analyze operator commands and sensor data to calculate the required output signals that drive actuators and motors to a desired end state. NI developed the roboRIO robot controller based on the needs of the FIRST® Robotics Competition (FRC®) for a rugged, reconfigurable, and reliable robot controller. The controller is integrated with a power distribution panel, a pneumatics control module, relays, voltage controllers, and sensors to form the complete robot control system. A wireless access point is used to communicate between the robot control system and the robot's driver. Typical robotics sensors include optical sensors, encoders, ultrasonic distance sensors, limit switches, gyros, accelerometers, and cameras. Five team profiles in this chapter offer examples of methodologies to monitor and control robotic systems.

Sensors, Monitoring, and Control Applications | 211

Team 624 – Sensing Victory

○ Automation was a key factor that influenced the design process to produce FRC Team 624's robot. The advanced features of the new robot control system were utilized to achieve superior performance.

COMBINING BASIC MACHINING AND ADVANCED SENSING

Cleverness and control were fundamental principles that guided the design and operation of *FIRST*® Robotics Competition (FRC®) Team 624's robot, Gravity. Cleverness was dictated by the team's need to design and manufacture its robot in rented space at a shopping center using portable basic machines. Control was realized by fully utilizing the features of the new FRC Control System released in 2015, including its ease of programming and the ability to incorporate a large number of sensors. Together these factors helped FRC Team 624, CRyptonite, from Katy, TX, produce a robot that won three regional competitions and was a *FIRST* Championship contender.

Iteration based on prototypes was also an effective strategy applied by CRyptonite. While the concept of prototyping is frequently applied to physical systems, FRC Team 624 also used this technique to optimize the performance of its control system. By developing prototypes of control functions and techniques, efficient methods to collect measurements were discovered and implemented. Data from a large number of sensors was then used to automate robot operations. By applying the prototyping process to both hardware and software, the same methodology was used to transform ideas to reality while simultaneously improving the overall performance of the designed system.

○ A mill and drill press were used to create prototypes and manufacture components that were installed on the robot.

○ A prototype design for the intake arm was modified to improve its performance and make better use of sensor data.

BEYOND THE BASICS

With its build site in rented space at a shopping center, CRyptonite was limited to basic machine tools for constructing its robot. All of the parts were manufactured by students, supervised by mentors, with a drill press, band saw, metal brake, lathe, and a small mill. The use of transportable manufacturing equipment developed into a strength as the limited machining options promoted creativity. The reliability of the manufactured components was enhanced by a three-stage design and manufacturing process that included the development of proof of concept prototypes, more precise prototypes, and, ultimately, competition-ready construction.

The goal of the proof of concept phase was to quickly develop prototypes to determine the most promising design candidates for additional iteration.

Wood was first used to create these prototypes for each of the robot's main systems. During this phase of the project, the less complex shop machinery such as the drill press and band saw were used with hand tools to manufacture components. To speed up the review process, previous robots were used as surrogate platforms to provide pneumatic power and motor control for the new prototypes. This phase provided a chance for inexperienced team members to develop manufacturing competencies using the most basic shop equipment and tools.

Metal became the material of choice during the second manufacturing phase as the wooden proof of concept prototypes become more refined and precise. A smaller and more skilled group of students was assigned to the manufacturing responsibilities during this phase of the project as more complicated shop tools such as the mill and lathe were required to build the more accurate systems. Because this was also an iterative phase it was important to limit the scope of work on any particular idea and avoid becoming committed to ideas that produced poor designs. Also, during this phase individual systems were integrated to create the team's practice robot.

Prototyping in this phase frequently led to "eureka moments" when creative solutions were discovered. For example, during this phase the ratcheting mechanism on the tote stacker was tensioned using nylon cord instead of aluminum — a change that was simpler, more effective, and lighter. This design was lower maintenance and minimized stress because the attachment point for the cord could be moved further away from the hinges.

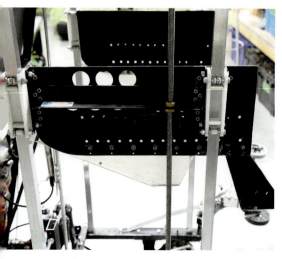

A rotating lead screw powered each side of the elevator carriage.

Also, during this phase a chain-in-tube system was refined to protect drive mechanisms from entanglement. This system was originally designed for the drive motors but later was also applied to lift totes. The high fidelity of the prototypes during this phase ensured the effectiveness of the improvements.

Progressing from prototype to becoming competition-ready was the third phase of the design process. This phase required the application of the most advanced machining techniques on the mill and lathe to fabricate components. Improvements to the practice robot were discovered and incorporated into the design of the competition-ready robot.

The elevator lift rails were fabricated from aluminum tubes, a material commonly available at any hardware store. A Delrin bracket held roller bearings against the tube to produce a low friction fit. The elevator was lifted by a rotating lead screw. A lower lip was added to the elevator to right containers that had fallen — a simple addition with significant impact. Testing of the system to ingest totes led to a realization that the intake wheels could be rotated in the opposite direction to move containers out of the way during the autonomous period. The use of existing systems for multiple purposes became creative solutions that were implemented with little cost.

ACHIEVING CONTROL WITH A NEW CONTROLLER

Scoring the maximum amount of points during the autonomous period was a team goal. FRC Team 624 rationalized that an investment in developing its control system would enable the team to excel during the autonomous period of play and benefit the team during the period when drivers controlled

A combination of sensors and mechanisms made it possible to pick up totes in any orientation.

The navX MXP Robotics Navigation Sensor provided a method to increase the number of sensors used on the robot. This board seamlessly integrated with the NI roboRIO robot controller.

○ A custom dashboard provided the robot operators with real-time feedback on all of the robot's systems and displayed data from all of the robot's sensors.

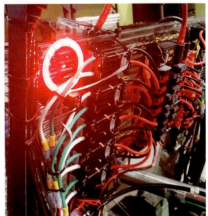

○ Sensors and signals included (from top to bottom) an encoder to measure elevator height, a limit switch to keep the elevator within its designed limit, a string potentiometer to monitor the stabilization system, an ultrasonic sensor to detect totes, and an LED signal light.

the robot. High levels of control would set CRyptonite apart from its competitors, further motivating the team to invest in this technology area.

FRC Team 624 benefitted from features incorporated in the new FRC Control System released for the 2015 season. Of particular value on the NI roboRIO robot controller was the myRIO Expansion Port (MXP) that enabled additional control functions. A navX MXP Robotics Navigation Sensor that provided a three-axis gyro sensor and included ports for additional sensors was installed on the team's MXP. This expansion port allowed for 15 digital input/outputs to be used in the control system.

The prototype design methodology applied in the build phase was also used to program the robot. Initial concepts progressed from a flowchart of the logic sequence to code that was initially tested on the optional AndyMark AM14U chassis – provided in the 2015 kit of parts – as a preliminary evaluation platform. Example code supplied with the controller was first used to learn about the new controller and to obtain data from the sensors. Other developmental advancements improved the control system. The dashboard program in the sample code was customized to display data most beneficial to the drivers. Also, FRC Team 624 modified the controller area network (CAN) Talon SRX LabVIEW library to increase the computational speed of this segment of the program.

A collection of sensors monitored robot operations. The drivetrain was outfitted with encoders to determine the distance each drive wheel traveled, a key parameter for autonomous sequencing. An encoder was also used to measure the height of the elevator with limit switches at the top and bottom of the run establishing the elevator's range. A string potentiometer measured the location of the stack stabilizer to keep this system within its intended range of travel. Ultrasonic sensors determined when totes were

◉ Each match began by raising the acquired tote and pushing the container out of the way while progressing to the second tote. The displaced container was kept upright to ease loading later in the match.

within reach and secured by the intake system. A digital pressure transducer measured the operating pressure in the pneumatics system and provided feedback that indicated the system's charge level. The gyro on the navX MXP Robotics Navigation Sensor provided the robot's heading and this signal was used to drive the robot in a straight path and execute exact turns.

GETTING A SENSE OF WHERE THINGS ARE

Control sequences using the collected sensor data were programmed to establish feedback signals that automated functions. The automated processes were faster and more accurate than driver-controlled maneuvers, thereby increasing the team's on-field effectiveness.

Automated solutions were developed for nearly every mechanism, allowing the drivers to focus less on individual processes and more on the game as a whole. As an example, an automated stacking sequence provided a very fast and accurate method for placing a large stack of totes on the scoring platform. In addition, included in the list of automated systems were: picking up different numbers of totes with various quantities of totes already in the elevator, depositing totes on the scoring platform, loading totes from the loading station, and righting fallen containers. Proportional-integral-derivative (PID) feedback loops were executed for those systems that needed to reach a defined set-point.

Because of the robot's integrated design there was a need to coordinate the operation of subsystems to avoid interferences between mechanisms. Safe operating ranges were established based on sensor data from the elevator, intake rollers, and stabilizer to ensure these components did not collide with each other. Other control applications eliminated potential system damage such as limit switches that provided a backup signal to prevent the elevator from exceeding its operating range. Also, motor current from the CAN bus detected when a motor was stalling, at which point it was stopped to prevent motor or actuator damage. The sensor data provided the needed input to program safe and smart robot functions.

Simultaneous integration of sensor data was essential to sequence functions during the autonomous period. This period of play began with the elevator limit switches establishing the lowest position of the elevator. This data was used to record the base value of the elevator encoder, with the sensor then measuring the height to raise the first tote to its correct position. This sequence was followed by the robot driving forward a prescribed distance based on feedback from the drive wheel encoders and the navX MXP Robotics Navigation Sensor, during which time the recycling container was pushed out of the way. This process was repeated to collect the three yellow totes. Having lifted the yellow totes, feedback from the gyro and drive wheel encoders commanded a precise turn into the auto zone, followed by lowering of the elevator and backing away from the scored stack.

RELIABILITY, CONTROLS, AND SUCCESS

The robot, control program, and strategy were adjusted many times over

The upper intake wheel was cleverly designed to upright fallen containers and ingest them into the robot.

The navX MXP Robotics Navigation Sensor helped process information and establish the timing for a series of distinct actions to acquire, transport, and deposit totes during the autonomous period.

◐ The team earned multiple awards at the Lone Star Regional, thanks to the robot's reliability scoring stacks of totes. The success resulted from a season of hard work and continual iteration by the team's build and control groups.

◐ Precise control and careful driving provided many opportunities to stack and score.

the course of the season to respond to newly discovered problems. Often the build and programming teams worked to develop hardware and software solutions simultaneously that improved the robot's on-field performance. One helpful feature was a light-emitting diode (LED) light ring on the robot that signaled when the robot was properly positioned at the loading station. This LED provided an unambiguous form of communication between the robot and robot drivers and minimized errors during intense moments of the competition.

The value of the team's ability to iterate hardware components and software functions was apparent in its performance. FRC Team 624 won three regional competitions, received design awards, and was one of the top teams at the *FIRST* Championship. This level of success was achieved using traditional machine tools and fundamental control algorithms — a combination of methodologies with applications well beyond robotics.

◐ A whiteboard was transformed into a giant flowchart to sequence events and sketch out programming algorithms to acquire and deposit totes.

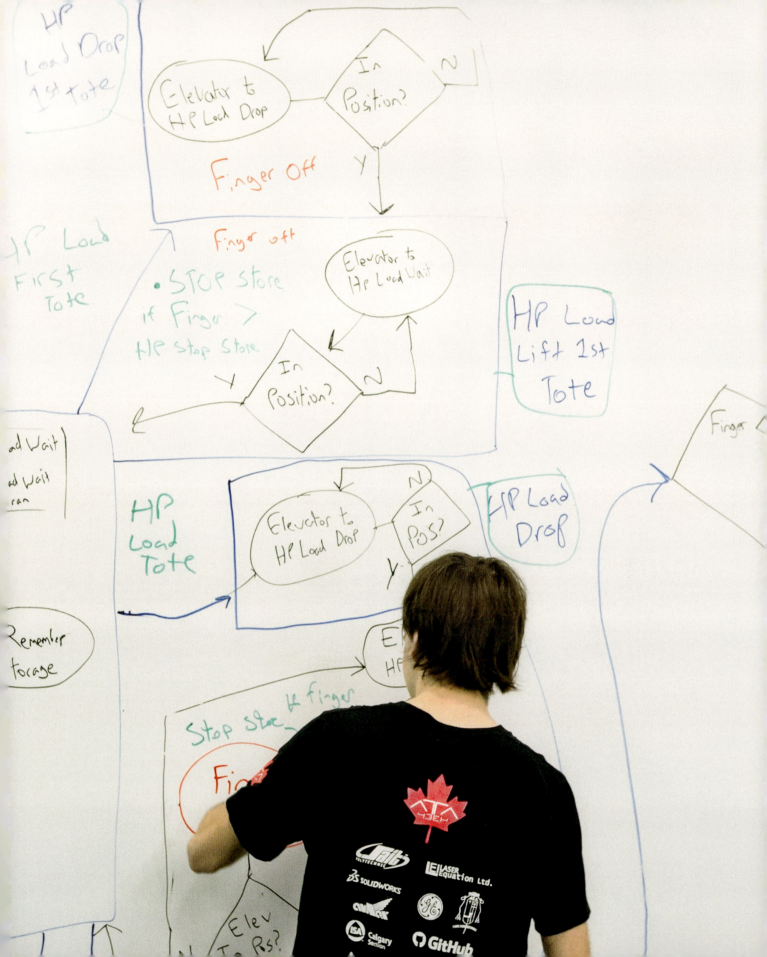

Team 1100 – Simply Reliable

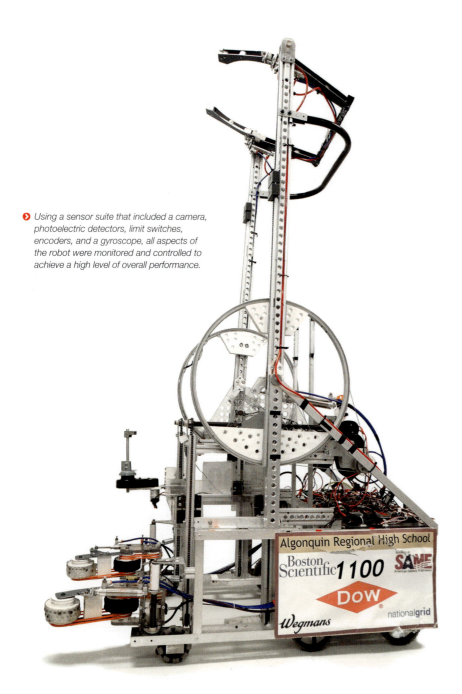

> Using a sensor suite that included a camera, photoelectric detectors, limit switches, encoders, and a gyroscope, all aspects of the robot were monitored and controlled to achieve a high level of overall performance.

EFFECTIVE AND EFFICIENT: HARDWARE AND SOFTWARE

The final rounds of a *FIRST* Robotics Competition (FRC®) are especially telling because the teams and their robots are battle-tested during one dozen qualifying matches. At this stage of the competition robots are often reaching their endurance limits given the number of previous matches. In addition, the frequency of play during the final competition rounds leaves little time for maintenance and improvement. The demands of near-continuous competition often establish the performance limits for specific robot systems once components wear and planned operations go astray.

For FRC Team 1100, the T-Hawks from Northboro, MA, robot fatigue during the elimination rounds was a non-issue. Throughout the competition, including the final rounds, each match for FRC Team 1100 started with the perfect execution of collecting three yellow totes and depositing them in the auto zone for an easy 20 points. During this process, the team's robot marked each autonomous round with a container toast to celebrate the team's planning, testing, and execution. Match after match, this flawless performance consistently met the team's initial design objectives. By emphasizing simplicity in design, effectiveness in sensing, and reliability in programming the team created a solid platform to compete and win.

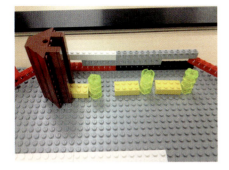

A small physical model of the playing field was constructed before developing a CAD model. The CAD model was used to explore the robot's interaction with game pieces.

The robot began the autonomous period ready to ingest a tote and lift a container. This action cleared the approach to the second autonomous tote, with an arm then rotating to move the second container out of the robot's path.

STARTING SIMPLY

Once the game details were announced, individual groups of students and mentors analyzed the challenge and created possible design solutions to meet the objectives. Considerable effort was devoted to this process, with three days spent in this phase, followed by presentations of the group work to the full team. The entire team carefully reviewed each group's design before combining the best elements from each to create the team's robot. Precision, speed, and efficiency were factors common to each group's proposal.

An overarching strategy crystallized during those first three days: maximize performance during the game's autonomous period. This strategy guided the robot design process and applied to the robot's configuration, operation, sensors, and control. The robot would need to automatically acquire three totes, move three containers out of its way, and follow a prescribed drive path to deposit the totes in the appropriate area of the field. Linked with this emphasis on designing around the autonomous period of play was a desire to apply simplicity as a design guide. Whenever a decision had to be made, the simplest solution became the preferred solution.

Guided by the emphasis on simplicity, an autonomous operation having the fewest number of maneuvers was desired. To accomplish the three-tote autonomous task, simplicity dictated that the robot move in a straight line to gather totes before turning and dropping the stack in the designated area. These factors required that the robot have a tote within its perimeter at the start of the match to enable it to move in a straight line and acquire the additional totes during the autonomous period. Given this constraint, a U-shaped configuration for the robot base was established. Each tote would be collected inside the robot and an elevator would lift the newly acquired tote, thereby opening up an area to gather a new tote.

Two other design features were derived from an emphasis on maximizing autonomous performance and instituting simplicity. Noting that the desired path forward to acquire autonomous totes was blocked by containers, two devices were created to eliminate these obstacles. An arm with a passive claw was included to simply lift the first container out of the way at the start of the match. This claw was hinged so that the

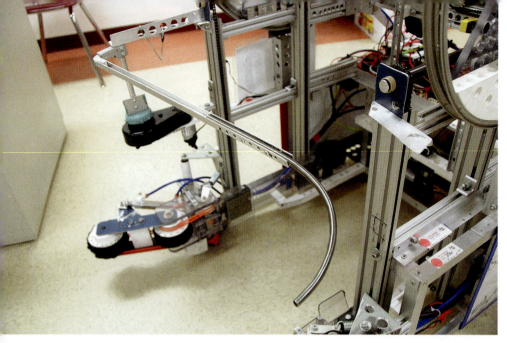

After knocking a container out of the way during the autonomous period, a limit switch measured the arm's location and signaled the arm's motor to stop, thereby protecting the arm from damage.

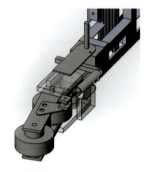

CAD models were used to design components and plan robot functions, including ingesting totes and lifting containers. The integration of sensors on the robot's structure was established at this early step of the design process.

distribution of the container's weight always kept the container in a vertical orientation. To move the other two containers out of the drive path, a single-action twitch arm rotated to knock each container out of the way.

The team's strategy for the teleoperated period was to acquire totes from the landfill zone. This strategy required that active intake rollers be included in the design, with these rollers actuated to maximize their effectiveness. As each tote was acquired, it was lifted by the elevator to create a stack.

Just as with the concept of simplicity guiding the robot's functions, this same concept was applied to manufacture the robot. The robot frame was constructed of one-inch aluminum angle having a thickness of 0.1875 inches, with the material riveted together at the corners. This frame was very light and provided minimal support. The elevator was fashioned from a combination of extruded aluminum bars measuring one inch by one inch and one inch by two inches. This structure was attached to the frame and produced a sturdy foundation to support the rest of the robot. Simple computer-aided design (CAD) models detailed each of the structural components used to construct the frame and elevator.

The emphasis on simplicity carried into the design and operation of the arm that lifted containers. Bicycle wheels were repurposed to become supports for the arm's base. A cable between a winch-driven take-up spool and each bicycle wheel produced the needed torque to lift the arm. Simple manufacturing and actuation methods were also used for the roller system where appropriately sized polyvinyl chloride (PVC) cylinders were machined to serve as drive pulleys and roller wheels. The rollers were pneumatically activated to facilitate the loading process, with a linear piston providing the force to open the intake.

SENSING AND SOFTWARE

A collection of sensors served as switches to record the position of robot mechanisms and ingested totes. Data from these sensors was used in both the autonomous and teleoperated

The intake mechanism combined motors, pneumatics, and tube-driven wheels to grip totes and funnel them into the robot. The wheels were mounted to a pivoting arm that maintained a secure grip on the totes.

periods of competition. The open-frame design allowed the sensors to be optimally placed throughout the robot to determine the location of totes and the orientation of components.

Autonomous code was written to achieve the team's design goal of maximizing scoring during this period of play. At the start of each match three parallel command sequences were executed. An initial sequence of commands lifted the first container to a maximum height to eliminate this obstruction from the robot's path. A photoelectric sensor was mounted on the robot frame. Reflective tape was strategically placed on the rotating bicycle wheel to activate the sensor and signal when the arm was vertical, prompting a command to stop the winch motor.

The second parallel sequence of commands drove the robot forward using the output from a gyro sensor to maintain straight line motion. During this drive sequence the twitch arm was activated to swat the second and third containers out of the way. A limit switch determined when the arm reached the desired angle of rotation, at which time the arm was restored to its home position and prepared to be reactivated. The robot's straight line motion continued until all three totes were loaded, with the gyro signal then used to execute a 90-degree turn. Following this turn, the robot drove forward and released the totes.

The third parallel sequence of commands controlled the intake rollers and elevator using two photoelectric sensors to detect the totes. A limit switch and an encoder that measured the number of times a shaft rotated were used to control the elevator. Photoelectric sensors mounted to the intake wheel mechanism detected the presence of a tote within the intake. This signal energized the intake motors to pull in the tote. This same signal activated the pneumatic cylinder to extend and clamp the tote with the powered forward pair of rollers.

A photoelectric sensor at the rear of the intake indicated that the tote was fully ingested. This signal prompted the control system to lower the elevator to pick up the tote, with the elevator's limit switch triggering when the elevator was at its lowest point and ready to lift the ingested tote. At this time an encoder on the elevator lift motor was zeroed to establish the base location for the elevator. The elevator was then lifted to a specified location based on the encoder's signal. The tote was lifted to a height that allowed a new tote to be ingested into the robot.

An optical sensor detected the presence of a tote in the intake. This signal triggered the piston to extend to more securely grasp the tote while it was ingested into the robot, thereby automating this process.

The arrangement of sensors on the robot signaled the presence of totes and the location of robot mechanisms.

The rotational position of the cable-driven rim that provided the torque to lift containers was measured using a photoelectric sensor.

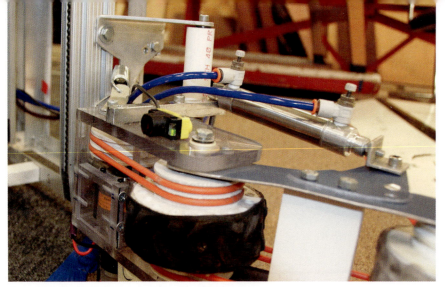

An optical sensor mounted above the intake wheel detected the presence of a tote, with this signal programmed to close the intake arms and establish a firm grip on the tote.

The control station consisted of joysticks, an Xbox controller, and a control dashboard that displayed a video feed from a camera mounted on the robot.

The process of picking up a tote based on feedback from two photoelectric sensors, a limit switch, and an encoder was repeated three times during the autonomous process. These signals also provided a passive indication of the robot's position on the field and triggered the initiation of the turning sequence once the third yellow tote was acquired.

The T-Hawks's investment in sensors was also a useful feature during teleoperation because the system to acquire totes functioned automatically during this period of play. The elevator system was augmented with a proportional-integral-derivative (PID) control algorithm to position any number of suspended totes at a constant height. A set-point was established to correspond to the desired lift height and the control system adjusted the power supplied to the lift motors to prevent downward drift due to the suspended weight.

A camera was added to extend the view of the robot drivers when picking up totes from the landfill zone. The camera and the automated intake system combined to create an effective method to position the robot and rapidly ingest and elevate the totes. The sensors were also helpful to deposit a stack of totes on the scoring platform. The simplicity of the robot design allowed the lifting mechanism to suspend the container under a growing stack of totes, with the stack of totes, in turn, lifting the container. As a passive device, the container arm did not require manipulation during scoring operations.

The team used Java to program the robot, starting with the sample code provided by NI as a framework for its coding process. With increased experience, the team's programmers progressed from modifying the supplied code to writing original code. The new FRC Control System released for the 2015 season simplified the testing process for tuning the robot code. The time required to test programs was decreased by an efficient debugging procedure embedded within the new control system.

TESTING ENSURES SUCCESS

Strategy and design simplicity were shared priorities, with the strategy defining the design envelope. The concept of design simplicity applied not only to the robot design, but also to the selection, function, and programming of sensors. Digital and analog switches established activation points for individual actions that were coordinated with input from the gyro sensor and the onboard camera. Together the suite of systems was reliable and effective.

The team's goal to maximize autonomous performance was achieved through rigorous testing and by executing its strategy. A competition field was constructed at the high school where the robot and its systems were tested, evaluated, and improved. The selection and placement of sensors, as well as the associated programming and debugging processes, were facilitated by the availability of resources to test the created systems. By emphasizing simplicity in design, the robot was completed in time to optimally tune the control system and ultimately achieve the team's goal of maximizing autonomous performance.

▶ Sensor data was used to coordinate functions, maneuver the robot, and reliably score points during the autonomous period. These sensors also provided valuable information to assist the robot operators.

Team 2062 – Value of Simplicity and Reliability

APPLICATIONS IN MECHANICAL AND CONTROL DESIGN

A simple system has an increased potential to be reliable. Achieving these paired performance factors usually demands that attention be devoted to each area while designing mechanical and control systems. *FIRST*® Robotics Competition (FRC®) Team 2062, CORE (Community of Robotics Engineers) from Waukesha, WI, emphasized simplicity and reliability while designing its mechanical and control systems, a strategy that served the team well during the 2015 competition season. Aided by simplicity and with a focus on the quality of its design, fabrication, and programming methods, CORE's robot performed with a high degree of reliability throughout the season.

Automatic control loops contributed significantly to the team's success, which included winning two regional competitions and participating in the *FIRST* Championship. The mechanical and control systems were well integrated with sensors acquiring performance and location data to control mechanical systems. The team used sensors and automatic control functions for a majority of the robot's mechanical systems to improve performance. By stressing simplicity in design and implementation, the systems achieved high degrees of reliability while accomplishing their required functions.

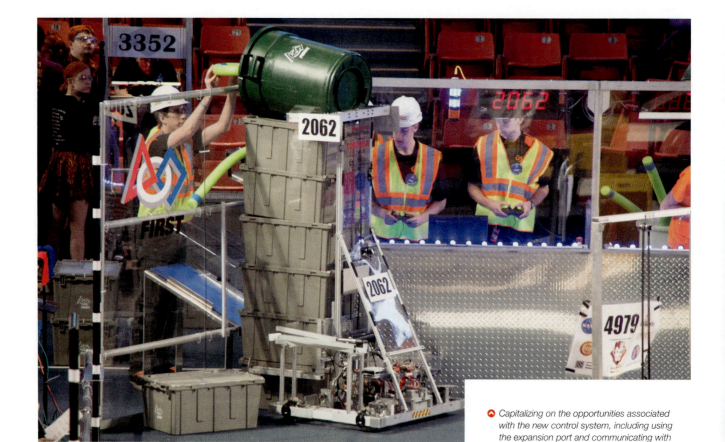

◉ *Capitalizing on the opportunities associated with the new control system, including using the expansion port and communicating with discrete controllers using the CAN bus, FRC Team 2062 mastered the ability to monitor and control their robot.*

226 | *FIRST* Robots: Behind the Design | Vince Wilczynski and Stephanie Slezycki

◆ Rotary encoders reliably measured the rotation of each drive wheel. These were essential measurements needed to control the mecanum drive system.

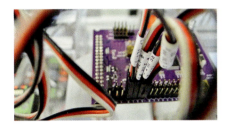

◆ The navX MXP Robotics Navigation Sensor provided three-axis accelerometer measurements and was a conduit for other sensor data.

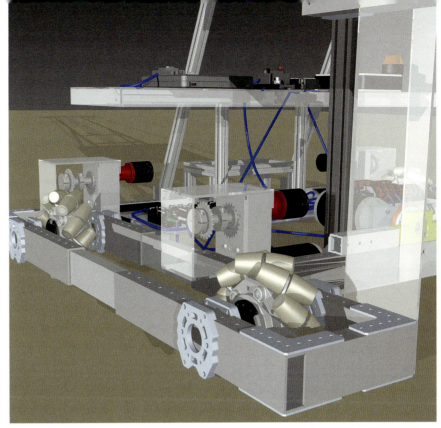

◆ Sensors, electrical panels, and control system components were included in the CAD drawings. This level of detail ensured that the electrical and control systems were integrated into the design process.

PROTOTYPING CONTROL SOLUTIONS

While all of the robot's mechanical systems were designed considering control aspects, two features in particular required special control functions: the drive system and the expanding chassis. Omnidirectional movement was achieved using mecanum wheels for the drive platform. To control the drive system, the power to each wheel was carefully regulated. Wheel rotation was monitored using rotary encoders installed on each transmission. These signals were combined with data from a gyro sensor on a navX MXP Robotics Navigation Sensor. This sensor connected to the myRIO Expansion Port (MXP) on the NI roboRIO robot controller. Data from all inputs was used to fine-tune the voltage for each drive system motor to achieve the desired directional and velocity control.

To increase stability while carrying a stack of totes, and meet the competition's transportation size requirements, the team designed a robot base with an expandable depth. This novel design produced a chassis that expanded six inches once the robot was placed on the competition field. The expansion was accomplished using rectangular Delrin beams that ran inside the four aluminum frame members. When the robot was on the field, the frame was extended and the Delrin inserts were locked to the aluminum beams. From a controls perspective, this design feature required flexible connections for the control and power systems located on the front of the robot.

The controls sub-team employed the concept of prototyping to build the control system. This process began with the construction of a prototype controls board equipped with all of the elements in the power, monitoring, and control systems. The proto-board provided students with an easily configurable platform to learn how the new control system functioned and to practice control methodologies. Signals from sensors and motor controllers were used to create feedback algorithms for each of the robot's functions. This board was temporarily fitted on the drive base as soon as it was constructed to quickly experiment with control code on the real drive system.

The entire robot was designed using Autodesk Inventor computer-aided design (CAD) software. One advantage of having the complete robot available as a CAD model was the team's ability to establish the electronics arrangement. This allowed the team to print a template for the controls platform, manufacture polycarbonate shelves for the control system, and wire the electronics components together once the components were fixed on the shelves. The modular control system was bench-tested before being installed on the robot.

SENSING SURROUNDINGS

Sensors, control algorithms, and mechanical systems were integrated to achieve desired control functions during both the autonomous and teleoperated competition periods. In addition to the drive system control algorithm, the tote acquisition and lifting systems also greatly benefitted from being automatically controlled.

To automate the process to acquire totes, three photoelectric sensors were mounted on the front of the robot, with the center sensor used to detect a tote in front of the robot. Once a tote was detected, the robot's drive system automatically maneuvered by twisting, strafing, and advancing to position the tote along the center of the robot. As the robot approached the tote, feedback signals from two other photoelectric sensors, each mounted on the far sides of the robot frame, were used to adjust the robot's orientation and keep the tote centered. This feature was especially useful to automatically align the robot at the loading station.

Ultrasonic sensors were mounted on the front and sides of the robot to measure the robot's location relative to other objects. Data from the front ultrasonic sensors was combined with input from the photoelectric sensors to carefully position the robot in the optimum location to receive totes at the loading station. Additional sensor input from one of the side mounted ultrasonic sensors controlled the distance between the robot and the player station wall. This set of sensor inputs provided multiple reference points to automatically position the robot on the field and optimize the robot's ability to acquire totes.

The robot's elevator was also automatically controlled based on input from a rotary encoder incorporated into the lift transmission. Four preset heights were defined in the computer

Rotary encoders helped control the drive system and monitor the elevator's height. Photoelectric and ultrasonic sensors were incorporated into the robot's base to determine the presence of totes and establish the robot's orientation.

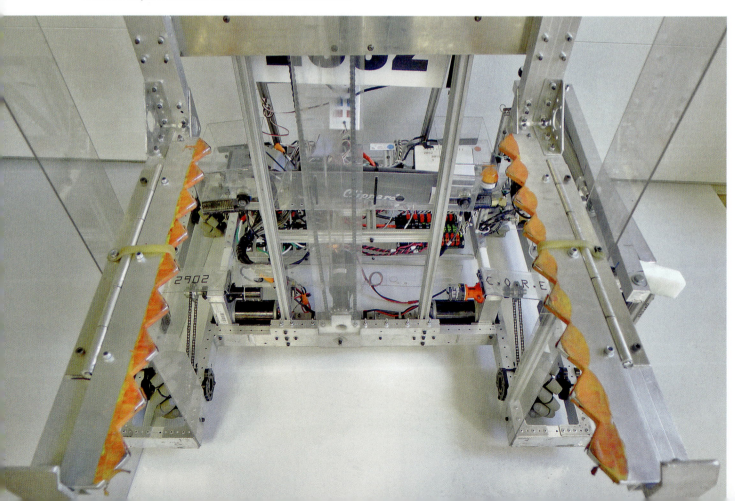

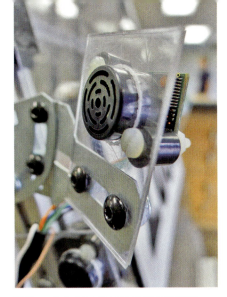

◊ One set of ultrasonic sensors measured the distance from the robot to the wall during the autonomous period, while a second set of sensors measured the distance between the robot and the tote chute.

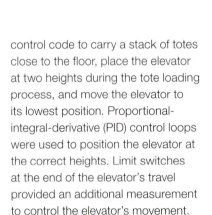

◊ A temporary control board was installed to test the robot's systems and evaluate the control system. The temporary mounting platform was later replaced with polycarbonate.

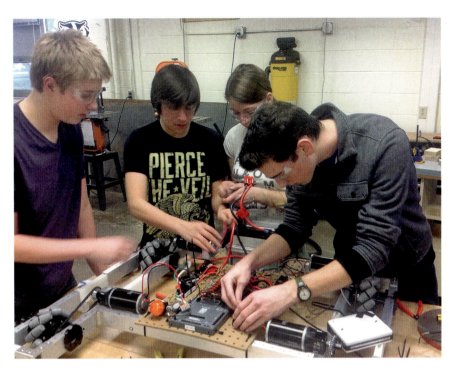

control code to carry a stack of totes close to the floor, place the elevator at two heights during the tote loading process, and move the elevator to its lowest position. Proportional-integral-derivative (PID) control loops were used to position the elevator at the correct heights. Limit switches at the end of the elevator's travel provided an additional measurement to control the elevator's movement.

The systems to align the robot and position the elevator were combined with the drive control system to operate the robot in a hybrid-control mode to quickly acquire a stack of totes at the loading station. Approaching the loading station, the driver would initiate this mode of control with the robot then automatically aligning itself to receive totes and then cycling through a series of steps to lift the collected load. This series of steps was faster and more reliable than having the robot operators execute a coordinated series of functions and movements.

MAKING THE MOST OF CONTROL OPPORTUNITIES

FRC Team 2062 took full advantage of the new control system introduced for the 2015 season. The span of supported programming languages in the new controller allowed the team to continue working in C++, the computer programming language they had used in previous competitions. The T-Hawks were able to benefit from its prior programming experience by using the training documentation to convert robot code from previous years into a format that was compatible with the new control system.

The team was also able to capitalize on its prior control programming work by importing a previously developed coding library into a format that could be used by the new control system.

Sample programming examples provided in the documentation for the new control system served as starting points to create custom-designed code for the 2015 season. As one example, the documented sample code for implementing PID control loops on the Talon SRX motor controllers helped the team quickly master this technique.

These programming advancements were not restricted to software. The roboRIO robot controller developed by NI had familiar features that helped the controls sub-team develop expertise with the new hardware. The roboRIO's expansion port was used to add a gyro sensor as well as other analog and digital sensors. The team made great use of the controller area network (CAN) on the new control system, with the CAN bus used to communicate with the motor controllers. Finally, the team made use of the hardware features on the

Signals from the ultrasonic sensors at the front of the robot helped position the robot at the loading station for optimal delivery of the totes.

Control code used the collected sensor data to determine robot functions, command motors, and energize actuators.

Talon SRX motor controllers, with the serial ports on these motor controllers used to provide feedback data for the embedded PID control loops.

INCORPORATING THE USER IN THE CONTROL PROCESS

The robot operators were also a key aspect of CORE's approach to control system design. The controls sub-team worked closely with the robot operators to garner their insights on functionality that improved the robot's performance. The performance of the drive system was one example of an improvement that resulted from this interaction. Creating a mechanically balanced omnidirectional drive system using mecanum wheels is challenging, given the many variables that affect each of the four propulsion sources. The programming sub-team and robot drivers communicated with each other to understand the particulars of the robot's drive system. With an understanding of the drive system behavior, computer code was written that corrected for the inherent mechanical differences and produced a reliable, robust, and maneuverable drive system.

Similar discussions and a joint awareness of the mechanical and control sequences to acquire totes also improved performance. To improve system reliability, programming code was developed that allowed the robot operators to calibrate control parameters for acquiring totes in real time during a competition match. If the pre-programmed default values for positioning the robot to acquire totes were incorrect, the operators could manually line up the robot during the match and save the location-pertinent sensor data from the ultrasonic and photoelectric sensors. The control code automatically accessed the new information and updated the control algorithm with the field-acquired data.

This level of improved operator-machine interaction improved accuracy and enabled FRC Team 2062 members to make the best use of their time on the field. The combination of simple and reliable robot hardware, programming software, and control functions resulted in the team winning two regional competitions, victories that would not have been possible without the team's special attention to sensors and control.

◗ Network cable jacks connected the ultrasonic sensors to the navX MXP Robotics Navigation Sensor to facilitate wire management. The connection of each sensor and motor signal to the robot's controller was an important detail that required careful and secure wiring methods.

Team 2168 – Custom Sensors, Software, and Driver Station Dashboard

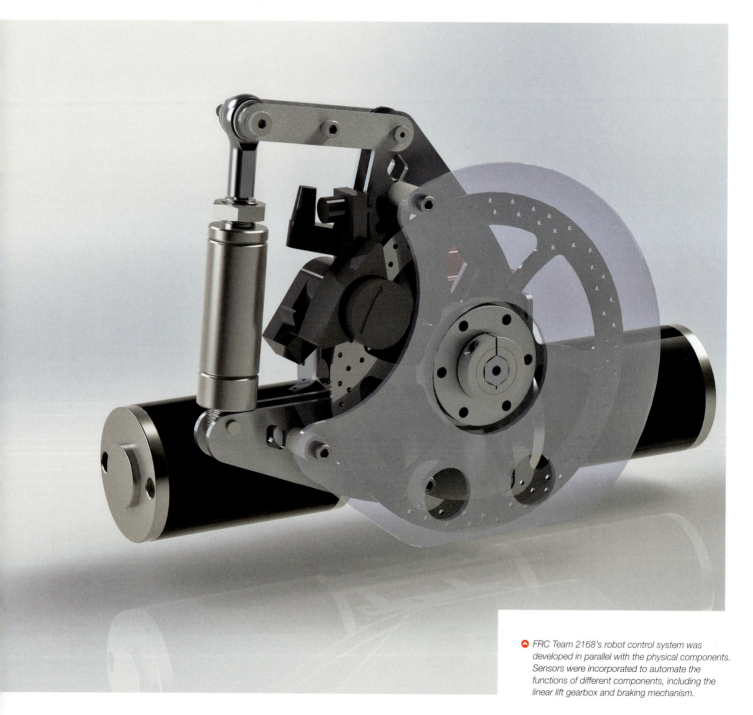

⬢ FRC Team 2168's robot control system was developed in parallel with the physical components. Sensors were incorporated to automate the functions of different components, including the linear lift gearbox and braking mechanism.

FIRST Robots: Behind the Design | Vince Wilczynski and Stephanie Slezycki

PARALLEL DEVELOPMENT OF THE CONTROL SYSTEM AND ROBOT

FIRST® Robotics Competition (FRC®) Team 2168, the Aluminum Falcons, knows how to control a robot. The team was selected as an alpha and beta test team for the new 2015 FRC Control System, and regularly supports other teams in the FRC community with robot programming and controls. For the 2015 competition, FRC Team 2168 incorporated optical encoders, Hall effect sensors, gyros, and infrared proximity sensors to automate key operations.

With almost one dozen pre-programmed autonomous modes to choose from and real-time dashboard diagnostics, this robot was an excellent performer. This Groton, CT, team emphasized an early focus on the design of the control system, which was developed in parallel, and sometimes ahead of, the actual robot components. Rather than the age-old chicken and egg question, the Aluminum Falcons introduced the question of which should come first: the control system or the robot.

MEET REGGIE, THE ALUMINUM FALCON

For RECYCLE RUSH℠, FRC Team 2168 set out to design and build a robot that was predominately focused on stacking. The machine, named Reggie, was divided into six fundamental subsystems: the drivetrain, a linear lift, a gripping mechanism for totes and recycling containers, an intake, a recycling container retention device, and a "bin ripper." The skid-steer drivetrain utilized traction wheels in the front of the robot and omnidirectional wheels with little sideways friction in the rear. As the robot maneuvered around the field, this configuration resulted in a turning point in the front, near the tote stack, minimizing lateral

◉ *Optical encoders provided feedback on the vertical position of the linear lift. Hall effect sensors, coupled with magnets, were used to detect limits of travel.*

movement of the stack while turning. The wheels were powered by two fixed speed transmissions, each with two MiniCIM motors. The linear lift was built to elevate the gripper and recycling container retention device. The vertical motion of these two attachments was driven by a belt and pulley system powered by two CIM motors with a worm gearbox. To prevent potential backdrive due to the heavy stacks, a pneumatic brake was added to the system.

The gripper consisted of two parallel arms actuated by pneumatic cylinders that opened and closed to lift totes and recycling containers. An intake mechanism, affixed to the chassis below the gripper, was incorporated to correctly orient and pull totes and recycling containers into the robot for the gripper to manipulate. The two arms of the intake had motor-driven wheels that pulled the game pieces in and reversed to push them out. The arms were pneumatically actuated to open and close around the game pieces. The recycling container retention device was positioned above the gripper on the linear lift to hold recycling containers while tote stacks were built below. Its vertical movement relied on the gripper and totes, and its arms were opened and closed through a spring return solenoid valve. The

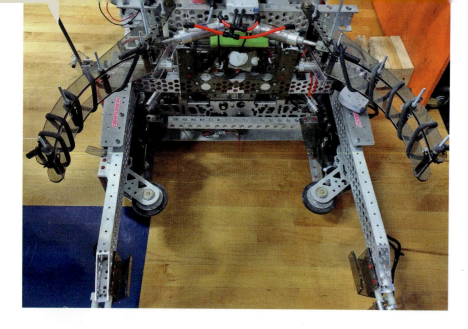

◗ Two motor-driven wheels were mounted on arms extending from the front of the chassis. This intake mechanism pulled game pieces into the robot.

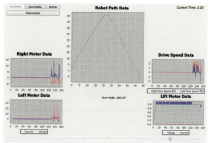

◗ A diagnostic tool was developed for reviewing logged performance data and performing trend analysis.

robot also had a "bin ripper" used to grab recycling containers off the step during the autonomous period. This system consisted of two large arms mounted on the back of the robot, actuated by pneumatic cylinders to collect two containers at a time.

CONTROL SYSTEM DESIGN, DEVELOPMENT, AND TESTING

The control system was integrated into the robot's design from the very beginning, rather than treated as an afterthought. The Aluminum Falcons used the Eclipse Java integrated development environment (IDE) to develop the robot code concurrently with the physical robot components. To begin generating a sequence of commands that would control the robot, team members identified the anticipated subsystems and interfaces early on in the design process, even though they didn't yet know how the components would function. Mainline code was tested as early as possible on robot prototypes to identify and mitigate potential problems. As the physical components were completed and assembled, the team began testing more complex code sequences. Despite the early testing, FRC Team 2168 continued to iterate the robot and software for improved performance and reliability throughout the season.

IMPRESSIVE AUTONOMOUS OPERATIONS

With nearly one dozen autonomous modes to choose from, FRC Team 2168's robot had the flexibility to adapt to its alliance partners' capabilities. The different options were selected by the drive team through a dropdown menu on the player station dashboard, and could be decided on right up to the start of the match.

The team's most challenging yet effective operation for the autonomous period was to stack the three yellow totes and move them into the auto zone. The programming to execute this mode consisted of 14 individual steps that required integration of a variety of sensors. At the start of the autonomous period, the robot's intake and gripper pulled in and lifted the first yellow tote, both intake wheels were driven clockwise (to sweep anything contacted to the right of the robot's path), the robot slowly drove forward until the recycling container had been pushed out of the way by the intake wheels, and the robot accelerated forward until it saw the next yellow tote. The robot then lowered the lift to engage the second tote, raised the lift to elevate the tote, again drove both intake wheels clockwise, drove forward slowly until past the recycling container, and accelerated forward until it saw the third tote. At this point in the sequence, the robot had two totes held by the gripper and one in the intake so it turned 90 degrees clockwise to face the auto zone. The robot drove forward into the zone, lowered the two elevated totes onto the bottom tote, disengaged the intake and gripper, and backed away from the stack.

This elaborate code evolved as the robot was developed, and the team spent the majority of the season perfecting it to run the full sequence in the short 15-second autonomous period. To reduce the time, a variety of solutions were considered, including driving the robot around the recycling containers, adding mechanisms to move the containers out of the way, and adding arms to drag the containers with the robot. After four separate iterations of the intake and the addition of forward-facing infrared sensors to detect the containers, the team was able to beat the 15-second goal.

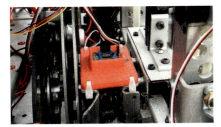

A digital output gyro, mounted onto a red three-dimensional-printed bracket, was used to control the direction of robot travel during autonomous mode.

Infrared proximity sensors were used to detect totes and recycling containers and control the spinning of the powered intake wheels to facilitate acquisition.

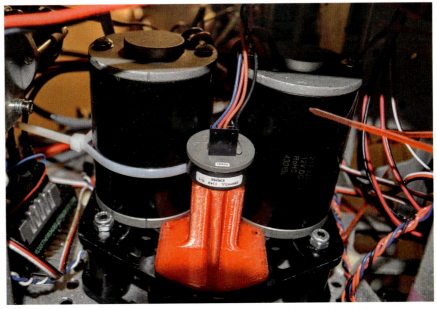

Optical encoders were mounted on three-dimensional-printed brackets and attached to the drivetrain gearboxes. The encoders provided feedback on the distance traveled by the robot and could control the distance driven when paired with a position controller.

SCORES OF SENSORS

Several types of sensors were incorporated into the robot design to automate some of the key functionalities. These automated sequences took the burden off of the drive team, and the attention to detail helped this team create a robot that reliably exhibited exceptional performance. Two Grayhill 63R256 optical encoders, one on each side of the drivetrain, provided feedback on the distance the robot traveled. When used within a position controller, the robot could be directed to drive a certain distance. The same type of encoder was used to provide feedback on the vertical position of the linear lift.

Two custom Hall effect sensors with light-emitting diodes (LEDs) that indicated power and sensed state were also used on the linear lift. Coupled with magnets, these sensors were used to detect the limits of travel without the need for contact of moving mechanisms. They were also used to zero the lift and prevent stall conditions if the lift was being operated in manual mode. An Analog Devices high performance, digital output gyro, the ADXRS453, was used to control the robot's heading in autonomous mode when it drove straight and then turned 90 degrees at a predetermined location. To interface with the gyro, the team wrote its own software class that included a recalibration function. This function automatically ran before the start of each match to monitor for errors and recalibrate if needed.

Three Sharp GP2Y0A21YK infrared (IR) proximity sensors were used to detect the game piece position relative to the robot to help automate the acquisition sequence. One IR sensor was positioned on the intake to control the spinning of the wheels so that the operator only had to press a single button to initiate the intake sequence. When a tote was being acquired, the intake arms opened to facilitate alignment. Once the tote passed a certain threshold, the intake arms closed so that the wheels would make contact with it. When the tote reached a second threshold, the wheels stopped spinning.

The other two IR sensors were used during the three-tote autonomous mode to detect the location of the recycling containers. One of the sensors was mounted to the front right chassis to detect when a container had been moved by the intake wheels so the robot knew to continue with the sequence. This sensor helped ensure that a container wasn't accidentally collected, and also indicated to the robot that it could accelerate forward once the container was out of the way. The robot couldn't drive fast while moving the containers because they would fall over instead of being pushed aside. The second IR sensor was mounted behind the linear lift, facing the front of the robot, to make sure that during the autonomous mode a tote stack wouldn't be lowered on top of a container if one

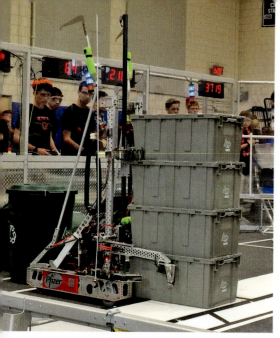

A custom dashboard provided the drive team with important information regarding real-time positions of the lift components, which was valuable when the view of the robot was obstructed by tote stacks.

happened to be pulled into the intake. This could cause stall conditions and snap drive belts, immobilizing the robot for the remainder of the match.

CUSTOM MONITORING AND DIAGNOSTIC SOFTWARE

Along with the various sensors, team members monitored and logged the current on every channel of the power distribution panel to which the motors were wired. One of the students developed custom software to review the logged match data and all sensor values. This data was reviewed with the corresponding match video to analyze the software and robot actions as a function of time. A trend analysis feature was used to review component performance over time to identify any hardware inefficiencies or failures. An additional custom software application was developed to monitor the electromechanical performance characteristics over time. Used as a diagnostic tool, this trend analysis helped identify imminent motor and sensor failures.

ENHANCING THE HUMAN EXPERIENCE

One part of the process that can be overlooked early on in the design phase is the layout of operator controls. Ideally, FRC Team 2168 wanted the most complex operations to be automated to ease the burden on the operator, while retaining manual controls in the event of sensor failures. The Aluminum Falcons developed a custom dashboard for the 2015 competition that was specialized to efficiently provide the drive team with important information. The upper right and left of the dashboard graphical user interface depicted two different virtual views of the robot that indicated the true position of each component based on the position sensed. This feature was used to verify the position of the linear lift, intake, and gripper. A quick glance at the dashboard provided the drive team with enough information to continue operating if their view became obstructed. This was especially useful with the tall capped stacks spread throughout the field.

Built-in diagnostic features were also added to the dashboard to quickly identify failure conditions. For example, the current consumed by each motor was represented by green, yellow, or red lights to alert the operators to potential motor stall. After each match, if robot performance problems were observed, the data could be analyzed to see if there were any correlating high current conditions.

KNOWLEDGE TRANSFER BENEFITS ALL FRC TEAMS

Since FRC Team 2168's first competition in 2007, the team members have come a long way in developing their skills in control systems and sensors. As an alpha and beta test team for the new 2015 FRC Control System, FRC Team 2168 helped review the system's functionality and provided feedback on bugs and incompatibilities within the software libraries and documentation. The team released documentation on computer vision software performance with the new system and compiled a resource of frequently asked questions for other teams.

FRC Team 2168 shares what it learned about controls and software with the FRC community and supports other teams with troubleshooting programming and control system problems. The team exhibited expert integration of various sensor types in its 2015 robot, and released design files for the custom Hall effect sensors, as well as source code for interfacing with the high performance gyro sensors. A real-time path planning application was also shared to provide guidance on complex autonomous robot motion.

The Aluminum Falcons operate on a set of values that include a commitment to excellence, pursuit of quality results, and empowerment of the students. The integration of the control system with preliminary robot design, impressive autonomous functionality, and custom sensors and software exist as evidence that these values are being practiced. The team's programming expertise and passion for educating others earned them the Innovation in Control Award sponsored by Rockwell Automation at the New England *FIRST* District Championship for the second consecutive year, and helped them to win the New England Pioneer Valley District Event.

> Preliminary robot design was performed in parallel with development of the control system. A SolidWorks model of the intake mechanism incorporated the location of a digital output gyro and optical encoder.

Team 3310 – Control of an Industrial Arm

A CONTROL FOCUS

Black Hawk Robotics, *FIRST*® Robotics Competition (FRC®) Team 3310 from Heath, TX, established two goals at the start of the competition season. The first goal was to design a robot capable of autonomously acquiring recycling containers at the start of each match. This action would tip the match in the team's favor by securing the additional containers for its alliance while simultaneously starving the opposing teams of scoring additional points. The second goal was to build a robot that sported a multi-axis industrial arm to manipulate totes and containers. The team recognized that this goal included two significant challenges: building the mechanism and controlling the team's creation.

The arm was modeled after a standard industrial robotic arm that is commonly used in automated assembly and inspection processes. The resulting design was a three-axis arm mounted on the robot base. The team's goals of autonomously acquiring the containers and operating a complicated multiple-joint arm could only be achieved using automatic control. One drive base, two container grabber arms, four independent joints and their respective motors, eight sensors, and 6,000 lines of code later, the team achieved its dual goals.

◐ With the ability to pivot at two joints and rotate around its base, the lifting system for FRC Team 3310's robot had the same range of motion as many commercial robot arms.

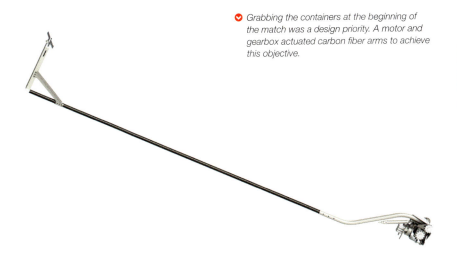

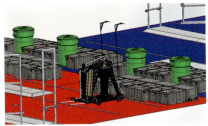

Grabbing the containers at the beginning of the match was a design priority. A motor and gearbox actuated carbon fiber arms to achieve this objective.

The container grabber was designed to avoid the landfill totes while reaching the containers. The geometry of the CAD model was confirmed with a physical prototype.

AN INDUSTRIAL ROBOT

The robot consisted of two primary systems: the device to acquire the center containers and the arm. The container-grabbing system consisted of two motor-driven lightweight carbon fiber arms that reached the containers as fast as possible. When the ends of the arms entered the holes in the container lids, small pneumatic pistons released barbs that engaged the lids. The containers were pulled away from the center step when the robot drove forward. This entire sequence was automatically executed and occurred in 0.25 seconds.

The three-axis robot arm consisted of a rotating turret, a lower arm, an upper arm, a passively controlled wrist, and a controlled gripper. A four-bar linkage was used to keep the gripper parallel to the floor during all arm movements. This ensured the stability of any object the gripper was holding. The turret was approximately three feet above the playing field floor. The robot drive system was powered using four motors with one motor for each of the robot's four wheels. A controllable ramp on the front of the robot guided totes from the loading station into the robot arm's reach.

SolidWorks software was used as a detailed design tool to establish the geometry of the arm and all other aspects of the design. The computer-aided design (CAD) for the robot was imported into the playing field CAD drawings to validate the dimensions of the designed systems. Dimension details were also verified with prototype testing. The electrical system was included in the earliest stages of component layout on the CAD model to incorporate this important aspect of the project into the preliminary plans.

The three joints for the turret, lower arm, and upper arm were powered by three independent motors and transmissions located at the base of the arm assembly. The power sources were placed at this location to minimize the height of the robot's center of gravity and to decrease the inertial forces that would result from a motor mounted on the upper arm. A mechanical linkage between the transmission and the upper arm's joint

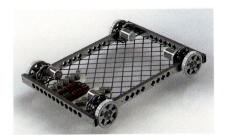

Each of the four drive wheels was independently powered. CAD was used to model and evaluate every component on the robot.

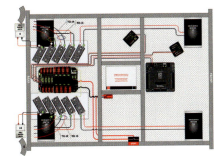

The layout of the electrical system was established to evenly distribute weight and have each electrical component easily accessible.

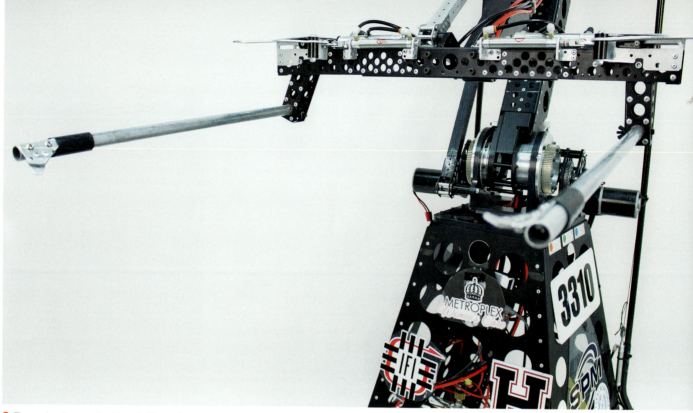

○ The arm's gripper evolved from holding a single tote to a device capable of holding two totes at a time, thereby doubling the robot's performance when stacking totes. The forked design was also capable of grabbing a container with the arm then placing the container on top of a stack of totes.

transferred the power to the upper joint. The tote gripper progressed from an early design that clamped down on one edge of a tote to a pair of pneumatically actuated rods that squeezed each side of a tote. This design revision allowed for two totes to be transported at the same time, thereby decreasing the cycle time to create stacks of totes.

A large number of sensors were needed to control all of the actions of the robot. Seven rotary encoders measured the angular rotation of the three arm joints, two of the four drive wheels, and the two container grabber arms. Optical encoders measured angular rotation on the wheels and joints. Miniature absolute magnetic shaft encoders on the container grabber arms determined their rotation. A gyro sensor recorded the robot's angle of motion, a key parameter to direct the robot's path. Given the importance of these sensors to control functions, encoders were incorporated in the CAD models for each system to best position the sensors to make the needed measurements.

PROGRAMMING THE AUTONOMOUS SEQUENCE

The carbon fiber arms of the container-grabbing system were designed to minimize weight and optimize deployment speed. An active control system executed a series of coordinated actions that began even before the match started. Proportional-integral-derivative (PID) control loops cycling at 1,000 times per second were activated on the Talon SRX motor controllers to control the deployment motors at the base of the arms. Though the motors were unpowered at the start of the match, the Talon's PID controller commanded that full power be delivered to these motors as soon as the match started. Using the internal control loop on the voltage controllers, full power to the deployment motors was triggered within one millisecond of the match's start.

The next step of the autonomous process was activating the robot's drive motors to pull the containers off the center step. If this action occurred too soon, the arms would miss the containers. If the action occurred too late, the opposing alliance would gain a time advantage to acquire the prized containers. Once the encoders mounted on the arm transmissions measured five degrees of movement, the robot drive motors were energized. With the arms moving at a high rate of speed and the drive motors needing to overcome their initial zero-inertia state, the motors were commanded to drive the robot before the arms even contacted the containers.

High speed video analysis of the system verified the timing of these coordinated actions. The short time window for these events required that the robot's control system operate at 200 cycles per second to provide the measurement resolution needed for coordinating these actions. Once the containers were acquired and pulled off the center step, the robot moved to a predefined location, released the containers, and navigated to the loading station for the start of the teleoperated portion of the match. The gyro sensor and drive system wheel encoders provided the needed data to move the robot along the defined path.

INDUSTRIAL CONTROL

The control algorithm for the three-axis arm enabled this system to automatically load, transport, and stack pairs of totes. This was accomplished using a variety of control methodologies including determining the required angular positions needed to achieve this motion, mapping the arm's path through three dimensions of space, and executing cascading inner and outer control loops while simultaneously coordinating the motion of each joint. The robot operator would initiate the automatic functions to pick up and deposit pairs of totes and then supervise the machine's actions as it automatically completed the assigned tasks. A seemingly simple operation was laden with subtleties and exacting details needed for success.

The power and speed of the roboRIO robot controller allowed for the computation of complex calculations to command each joint to move to prescribed locations. Using the joint angles and the geometry of the robot's base and two arm sections, the location of the end effector was calculated as X, Y, and Z Cartesian coordinates in physical space. This process of solving the forward kinematics equations was completed 50 times per second by the controller to determine the end effector's location. These results were then applied to determine the joint angles needed to position the end effector at a specific location.

Motion planning was used to establish a path of motion for each joint at every point in time to coordinate the movement of the robot arm. By solving the inverse kinematics equations, the joint angles and speeds for each point in the motion plan were determined. This series of joint positions produced the arm's coordinated motion. The calculations were completed once per computing cycle, placing a significant data processing load on the robot controller.

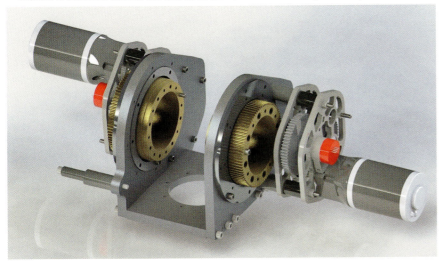

A separate motor and transmission, located on a rotating turret, powered each joint. Encoders provided feedback to the control software that allowed the arm to move autonomously from one position to another.

The container grabbers reached the center containers in approximately a quarter of a second, making this design one of the fastest devices at each competition.

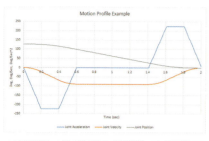

○ Motion profile programming, commonly used with industrial robots, established a series of location waypoints for the joints to pass through. Based on each joint's maximum velocity and acceleration, the control algorithm determined the arm's optimal motion profile.

○ Hollowed out gears transferred the transmission torque to actuators that produced angular motion. A linkage rotated the upper joint, and this mechanism enabled the upper joint motor to be mounted at the base of the turret.

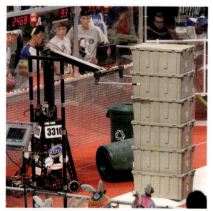

○ Parked in front of the loading station, the robot collected two totes at a time then stacked each pair on the scoring platform.

The robot controller was programmed in Java and augmented with a modified version of the WPI Robotics Library (WPILib) command-based architecture. The controller area network (CAN) bus sent and received signals from the control components. All sensors were connected directly to the Talon SRX motor controllers that executed internal PID control loops. Running PID control loops directly on the motor controllers distributed the overall computational workload, allowed for cascaded control loops on the motor controllers and roboRIO, and preserved the roboRIO for the computationally demanding forward and inverse kinematics calculations. The web-based interface for the roboRIO and the CAN-enabled devices was key to the team's effectiveness as it facilitated rapid setup, debugging, and control system tuning.

Control of the robot arm required a precise orchestration of movements to coordinate motion, avoid obstacles, and reliably deliver the totes to the designated positions. A motion profile planning technique, commonly used to program industrial robots, was used to determine all aspects of the coordinated motion of each joint. With this technique the physics of the robot arm were coupled with the commanded acceleration and velocity of each joint to create a smooth motion path for the arm as a function of time. The team's computer code implemented PID control loops to follow the points along the path for each joint.

Motion profile planning identified a list of waypoints for the arm to pass through, as well as the maximum velocities and accelerations of each joint. These waypoints were specified

The long reach of the arm allowed the robot to grab containers from the center of the field and deposit the container on top of a stack of totes.

as the Cartesian coordinates for the arm's end effector at specific moments in time. The computerized planner used this information to calculate each joint angle at each point in time to achieve the desired motion. The profile planning code also determined the acceleration and velocity for each joint to obtain the optimal motion profiles. Specifying a percentage of the maximum velocity for each path segment modified the arm's speed of motion. As an example of the amount of data computed, the motion profile to move a stack of totes from the loading station to the scoring platform was defined by six waypoints for each joint during the six seconds this operation required.

This motion was coordinated with a sequence of steps to grip the totes, move the arm, release the totes, pause, and return the arm to the loading station. While the operator was capable of independently executing each operation, the sequence was coded into the control algorithm to be automatically executed. The operator would select specific tasks for the robot to execute, such as loading and placing totes, and then allow the controller to automatically complete the tasks.

GETTING THE SYSTEM UNDER CONTROL

The robot was designed from the beginning knowing that an automatic control system would be required to have the system function properly. To prepare for this requirement, a control system station was constructed to investigate control details before the robot was available. The forward and inverse kinematic equations were also developed before the robot design was completed, with the specific parameters of the joint locations and arm lengths incorporated into the computational model once the robot configuration was confirmed.

Even with this advanced preparation, most of the evaluation and tuning of the control program was accomplished using the completed robot. This evaluation began with establishing the velocity and position control loops

◆ Laying out the wiring and electronics in CAD before the robot was built decreased the time needed to wire the robot.

for the drivetrain. Once control of the drive system was stable, testing of the robot arm began. This testing was accomplished one axis at a time to confirm the sensor feedback and establish operating limits for each section of the robot arm. The gains for the controller were then determined to achieve the desired motion.

Code development for automating the loading and stacking processes progressed through the competition season. During early competitions, the operator manually controlled the arm location and the end effector operating sequence. By the FIRST Championship event the entire tote loading, transport, and depositing process was automated to save time, increase reliability, and improve performance. In addition to its on-field performance, receiving the Innovation in Control Award in the FIRST Championship Newton Subdivision signaled that FRC Team 3310 had its robot under control. By successfully integrating industrial robot arms and industrial approaches to programming, the team showcased this technology and met its dual goals of speed and control.

◆ Four-bar linkages rotated the wrist and the upper arm. CAD models confirmed the length of each linkage to achieve the desired extension.

244 | FIRST Robots: Behind the Design | Vince Wilczynski and Stephanie Slezycki

◗ *Performance specifications and kinematic dimensions were combined with output from seven rotary encoders to precisely control the robot's industrial-inspired arm.*

★ AFTERWORD
FIRST® — A Community Working as One

This book illustrates technical excellence. It features engineering and designs that teach hard-sought lessons about solving complex hardware and software problems. That excellence is the product of a complex dance. People, individually and in teams, met Mother Nature's laws and they learned to rock and roll together.

The *FIRST*® community is a fractal set of problem solvers. Individuals work with nature's resources and within nature's constraints to design and build. And, the problems that teams solve are not limited to the technical domain. Teams of individuals work together to raise funds, work out schedules, and help their communities. The community of teams pushes toward a goal of changing culture. Each layer of the *FIRST* network works in an iterative and creative cycle.

The *FIRST* community produces thousands of "solutions" to the same "problems." One rarely sees that. We can learn a lot by studying all of the many paths others took.

Each CAD file, each chunk of metal, each block of code, and each innovation in team function includes the DNA of overall success. *FIRST* is a nested, fractal, and creative community of very nice people.

Members of the *FIRST* community are gracious professionals. They treat one another with respect and value the contributions of all. They have fun and get very tired working hard on tough problems. They celebrate learning and knowing. They are enthusiastically optimistic and honestly realistic. They have empathetic responses to one another.

The delightful designs in the book are the result of years of evolution. Each particular innovation may be new, but its inventors got to stand on the shoulders of many others. This enabling started a few thousand years ago, but was very heavily influenced by the *FIRST* community's amazing 25 year history. That history laid the foundation for learning to profit from "failure," for allowing technology to be cool, and for sincerely celebrating others' successes.

And, the work of each *FIRST* team today becomes a part of the DNA of every *FIRST* team in the future.

Dr. Woodie Flowers is the Pappalardo Professor Emeritus of Mechanical Engineering at the Massachusetts Institute of Technology and a Distinguished Partner at Olin College. He serves as Distinguished Advisor to FIRST®, as well as Co-Chair of the FIRST Executive Advisory Board, and participated in the design of the FIRST Robotics Competition game for many years.